The Origin of the Idea of Chance in Children

Also by Jean Piaget

The Child's Conception of Number
Genetic Epistemology
The Origins of Intelligence in Children
Play, Dreams and Imitation in Childhood

with R. Garcia

Understanding Causality

with Bärbel Inhelder

The Child's Conception of Space
The Early Growth of Logic in the Child

contributor

Play and Development
(edited by Maria W. Piers)

JEAN PIAGET
and BÄRBEL INHELDER

The Origin of the Idea of Chance in Children

Translated by Lowell Leake, Jr., Paul Burrell, and Harold D. Fishbein

The Norton Library
W·W·NORTON & COMPANY·INC·
NEW YORK

Originally published in France as *La genèse de l'idée de hasard chez l'enfant*. Copyright 1951 by Presses Universitaires de France

First published in the Norton Library 1976

Published simultaneously in Canada
by George J. McLeod Limited, Toronto

Library of Congress Cataloging in Publication Data
Piaget, Jean, 1896–
The origin of the idea of chance in children.
Translation of La genèse de l'idée de hasard chez l'enfant.
"Major related works by Piaget and his collaborators:"
p.
1. Child study. 2. Chance. I. Inhelder, Bärbel, joint author. II. Title.
BF723.C4P513 155.4'13 74–11040
ISBN 0 393 01113 5
ISBN 0 393 00803 7 (Paper Edition)

PRINTED IN THE UNITED STATES OF AMERICA

1 2 3 4 5 6 7 8 9 0

Contents

Part Three: Combinatoric Operations

Translators' Preface

Jean Piaget, born in 1896, is an intellectual giant in the field of developmental psychology. Although he earned his doctoral degree in the natural sciences (primarily biology) in 1918, his knowledge ranges in great depth through philosopy, religion, sociology, logic, mathematics, and, of course, psychology. Currently, the major emphases of his research and writings are on perception and memory experiments and on a complete analysis of the problems of genetic epistemology. During the past ten to fifteen years, the prolific contributions of Piaget and his collaborators, with Bärbel Inhelder being the most prominent and important of these, have attracted the rapt attention of psychologists and educators in the United States. His name is now well-known here, but the extent of his contributions is less well-known. John Flavell [1] in 1963 mentioned that more than twenty-five major books and over one hundred fifty articles exist for scholars to examine; few of the articles have been translated and not all of the books.

This work, *The Origin of the Idea of Chance in Children* fits in with earlier works concerning quantity, logic, number, time, movement and velocity, space, geometry, and adolescent reasoning. Some of these works are included in the Bibliography, but Flavell's book offers a comprehensive listing of works up to about 1962. Needless to say, many more publications have appeared since then.

The Origin of the Idea of Chance in Children was first

[1] John H. Flavell, *The Developmental Psychology of Jean Piaget* (Princeton, N. J.: Van Nostrand, 1963).

published in 1951 in Paris, France, but it has remained mysteriously untranslated until now. This twenty-three-year lag seems even more remarkable since the work is a major one by Piaget and Inhelder. Furthermore, translations of other early major works appeared as long ago as 1926 (*The Language and Thought of the Child*) and as recently as 1970 (*The Child's Conception of Movement and Speed*). And finally, this lack of a translation until now seems particularly strange since the work deals with chance and probability, mathematical concepts which have come to play a more and more important role in the mathematics curriculum of our schools, from kindergarten through graduate university education.

Hopefully this translation will provide interested psychologists and mathematics educators with a new catalyst for research in cognitive development. It should also prove useful to the mathematics teacher at the precollegiate level who has an interest in the theory of cognitive development and who teaches probability in the classroom. It is, in fact, one of the few works by Piaget that the beginner can read profitably, without needing too much help from the critics, because it gives a relatively clear example of the development of his techniques and thought. Flavell expressed the warning, "Furthermore, most of Piaget's writings are very difficult to read and understand, in French or in English. For one thing, there are many new and unfamiliar theoretical concepts, and they intertwine with one another in complicated ways to make the total theoretical structure. In addition, much of the theoretical content requires some sophistication in mathematics, logic, and epistemology." [1] Happily, Flavell softens this caveat somewhat by saying "Fortunately, *The Genesis of the Idea of Chance in Children* is a fairly easy book to read (as Piaget books go), and its concluding chapter offers an excellent summary of the principal findings and Piaget's interpretation of them." [2] Most readers of any of Piaget's works would do well to read Professor Flavell's book in conjunction with reading Piaget's own books; Piaget himself recommends Flavell's book in its Foreword, and it seems to be the single, most sophisticated analysis of Piaget's theories available in English.

[1] Ibid., p. 11.
[2] Ibid., p. 341.

Foreword

The present work is a supplement to our previous publications on the development of thought in the child and is the result of two preoccupations.

After having studied how logical, mathematical, and physical operations develop in the mind of the child and are adapted to that part of his experience which can be structured deductively, our next question was how thought which is in the process of formation acts to assimilate those aspects of experience which cannot be assimilated deductively—for example, the fortuitous or the randomly mixed.

Our second concern is that an analysis of intellectual operations and related phenomena in no way exhausts the genetic functions of the thought of the child (and still less that of the adolescent). What remains to be discovered, therefore, is how the mental processes work in the totality of spontaneous and experimental searchings which make up what is commonly called the problem of induction. Induction is first, and perhaps foremost, an effort at sifting our experiences to discover what depends on regularity, what on law, and what part remains outside these—i.e., what is simply chance. As an introduction to our investigations of intuition, we have recognized that a study of the origins of the idea of chance is unavoidable.

We are presenting here from this double perspective the results of some experiments on the psychology of the idea of chance. The reader will find in these pages material complementary to the operational analysis of the thought of the child; he will find, in addition, a sketch of further possible research into the formation of experimental induction.

Jean Piaget
Bärbel Inhelder

Introduction

The Intuition of Probabilities

A mathematician known for his work on probability theory suggested to us one day a study of the following problem: Could there be in a normal man an intuition of probability just as fundamental and just as frequently used as, say, the intuition of whole numbers? Almost every common action seems, in fact, to require the notion of chance as well as a sort of spontaneous estimate of the more or less probable character of feared or expected events. We know, for example, that there is a better chance of finding an object lost in a small space than in a large one. To avoid getting hit while crossing a street, we are making judgments every moment about the speed and position of the cars. After noticing a halo around the sun, if it rains the next day, we will say that it is not chance, while on the other hand rain on three Sundays in a row will not make us conclude any sort of natural law concerning rain on Sundays. We recognize, finally, that the inextricable mixture of facts and causal sequences forces us to take a probabilist attitude. It is only in theory, and at times in the laboratory, that phenomena are simplified to the point of allowing for the formation of a particular hypothesis. In our daily lives, all occurrences are complex: The fantastic path of a falling leaf is seen more often than is a straight line, and this is why we are reduced all our lives to guessing or to basing our expectancies on empirical frequencies and on contingencies.

But if we cannot deny that there is an intuition of probability in the normal civilized adult, and if we can correctly compare the role of this intuition to that of several practical operations such as number and space, there are nevertheless two questions which must be asked at the start: Is such an intuition in-born or does it develop later and, if so, how is it acquired?

Leaving aside both those psychopathological states in which the notion of chance is lost in a welter of obsessional or frenzied interpretations which load the most fortuitous events with subjective meanings and also those states of passion of a lover or gambler which are characterized by a like regression in favor of symbolism or a play of imaginary tokens, there are still two perfectly normal psychological states in which the understanding of chance and probability seem more or less absent: the primitive mind and the mind of a small child.

We are well aware that M. Levy-Bruhl considered the absence of the idea of chance an essential characteristic of the primitive mind. Since the primitive saw every event as the result of hidden as well as visible causes, and since he lacked the rational or experimental criteria to rule out even the strangest and most unforeseen connections, the prescientific mind could not have an intuition of probability as we have. The modern idea of chance is contrary to both types of causality, to determinism and to miracles. On the one hand, it differs from purely mechanical determinism, whose spatial and temporal links in the ideal state are reversible, because probability implies the intervention of irreversible phenomena. On the other hand, by interference within causal series, probability excludes categorically the concept of miracle. It assumes precisely that this mixture has its laws, while a miracle is the negation of such laws. In considering the primitive mind, a further question is, At what point does primitive thought perceive the possibility of mechanical causality? The question of miracle, however, cannot even be asked, since anything for the primitive can be a miracle. Here then are two reasons why the primitive remains innocent of the idea of chance, certainly to a greater degree than we do.

To use the primitive mind solely, however, in our analysis of the genesis of these notions leaves us uneasy. Levy-Bruhl's works have shown brilliantly the ideological and even mythic nature of primitive collective concepts. The whole technical side as well as the individual's daily differentiated use of these primitive concepts escape us still. In our examination of the idea of chance these aspects of their thought have an importance. We know quite well how the primitive attributes death, sickness, accident, and misfortune to the intervention of hidden powers and excludes the idea of chance. But we would certainly like more information on how the Arunta or the Bororo goes about finding a misplaced tool, or how he reacts when, taking aim, he is caught both by those kinematic laws which control the flight of his arrows and also by the fortuitous arrangement of things around his target.

It is at this point that our observations of the child are relevant. Certainly the mind of the child is always dependent on the surroundings, and each person, therefore, moves through an ensemble of collective representations in his development which the family and the school impose. But this circumstance is far from being a hindrance to our particular problem in psychogenetic studies. Suppose, for example, that in spite of the fact that intuitions of chance and probability are an integral part of our society's common sense, a small child, even one raised in an intellectual family, were resistant to such notions up to a certain age. This would prove first that mental operations are not entirely dependent on the collective milieu, and secondly, would permit us to analyze precisely how the idea of chance develops.

Not only do we find that things happen this way, but we find also that the formation of the rational processes in the mental evolution of the child throws light on the nature and conditions of the genesis of the ideas of chance and elementary probability.

It is quite natural that the child does not have at the very beginning an idea of chance, because he must first construct a system of consequences, such as position and displacement, before he would be able to grasp the possibility of the interference of causal series or of the mixture of moving objects. This development of the idea of causality and of order in general assumes an attitude exactly opposite to the attitude which can recognize the contingent and the fortuitous. Thus the idea of chance and the intuition of probability constitute almost without a doubt secondary and derived realities, dependent precisely on the search for order and its causes. As we have shown elsewhere, this is seen when we examine spontaneous questions of children, especially the famous Why questions which adults have so much trouble answering. The Why is asking the reason for things in cases where a reason exists, but also quite often in cases where it does not; that is, in cases where the phenomenon is fortuitous but where the child sees a hidden cause. Questions such as the following are pseudo questions: Why doesn't Lake Geneva go all the way to Berne? Why is there a Big and a Little Salève? Why isn't there a spring in our garden? Why is this stick taller than you? Why are you so tall and yet have small ears? And so on. All of these are pseudo questions for us because the facts to be explained are all due to chance interferences in biology and geology, to chance encounters, and the like, while the child who supposes that there are reasons for everything asks for the reasons in exactly those cases where they are least apparent. He has not yet understood that these are precisely the cases where there are none.

And there is more to it. If the study of the thought of the child brings us to recognize that the idea of chance and the intuition of probability are not innate, not primitive, then the psychogenesis of these notions during mental development throws a particular light on the formation of these ideas which have become fundamental in contemporary scientific thought. One particular aspect of the intellectual development of the child is to establish a progression between irreversible actions (both motor and perceptive) and rational operations, actions which become reversible and interrelated. The central problem which the evolution of intelligence in each individual permits us to resolve is the question of the genesis and nature of logical and mathematical operations in so far as these operations derive from experience and are structured by a reversible process. From the prelogic of the child, which is characterized by irreversibility, to primitive modes of thought, to the beginnings of logical-mathematical reasoning, one can follow step by step, through the course of a dozen years or so, the mechanics of this sort of development of the human mind. And the essential aspect of this development of the individual's thought processes depends on his gradual recognition of the reversibility of operations.

One can see immediately the interest that this situation holds for the ideas of chance and probability. From the physical point of view, chance is an essential characteristic of irreversible mix, while at the other extreme is mechanical causality, characterized by its intrinsic reversibility. Could it not be then that the discovery of the idea of chance, that is to say, the very understanding of irreversibility, had to come after the understanding of the reversible operations, since we mean by chance that part of the phenomena which is not reducible to reversible operations? And are these not the only operations which will allow us to grasp their opposite? This means then that thought which is still irreversible, that is, thought which moves in a single direction and which is completely dominated by the temporal course of events, this thought is not touched by the idea of chance precisely because it lacks the mental organization capable of distinguishing reversible mental operations from fortuitous events.

In a word then we must visualize chance as a domain complementary to the area in which logic works and, therefore, not to be understood until reversible operations are understood and then only by comparison with them. In such a case probability would be a counterpart of mental operations; that is, an assimilation of chance with the combining operations. Since we are unable quite simply to deduce each interference, it is the mixture taken as a whole which the mind assimilates. After this complex operation we have the reduction of the particular cases to all of their possible combinations.

These are the hypotheses our study has yielded on the development of the ideas of chance and probability at their most elementary stages—that is, in children from the age of four or five to eleven or twelve. The usefulness of this study lies not in any theoretical analysis of our data; rather, it has seemed to us necessary to make the facts themselves available in the order of their appearance. The facts will tell us then as we examine the correlative development of operations, of the idea of chance as their opposite, and of the probabilistic assimilation of chance with the combinatoric mechanisms.

In the first part of this work we will examine the formation of the physical aspects of the notion of chance: that is, the idea of an irreversible mixture, of diverse distributions (uniform or centered), which characterize fortuitous events, and the relationship between chance and induction.

The second part will examine one group of random subjects of various ages and another group of special subjects. At this point we will move from the interpretation of chance to an examination of the beginnings of the quantification of probabilities.

Finally, a third part will consist of the analysis of the development of combining operations—combinations, permutations, and arrangements. The juxtaposition of this development with the preceding facts will allow us to determine, as a conclusion, the relationship between chance, probability, and the operating mechanisms of the mind.

PART ONE

Chance in Physical Reality

I. Notions of Random Mixture and Irreversibility

It is quite probable that the concept of chance starts from the idea of an increasing and irreversible combination of phenomena. Cournot's famous interpretation conceived of physical chance as the interaction of independent causal series. But this complex notion, which needed an understanding of both interaction and independence, could hardly be grasped by the uninformed mind except in those cases where an intentionalist interpretation was eliminated because of the large number of elements involved. When a child is struck by a door which a gust of wind has closed, the child will find it difficult to believe that neither the wind nor the door had the intent of hurting him; he will certainly see the interaction of causes which brought him near the door, and also what caused the door to move, but he will not admit their independence. It is this fact which will not let him see the event as fortuitous. On the other hand, he does not recognize that chance characterizes daily happenings (social, meteorological, etc.) because he fails to notice the interactions of phenomena (e.g., the relationship between night frosts and the flowering of a fruit tree). In brief, the alternatives which have kept the child (as well as the primitive mind) from constructing the idea of chance are his recognition of either an interaction of causes with no recognition of their independence, or their independence without realizing their interaction. On the other hand, a combination of a sufficiently large number of elements seems to give a situation favorable to the intuition of causal series which both interact and are independent since the sequence of events is easily established with no need for imagining that there is anything intentional in the details.

The problem then is to determine if the child, in the presence of an obvious mixture of material objects, will perceive it as an increasing and irreversible mixture of the objects; or if, in spite of the obvious disorder, he will imagine the different objects as still being linked by invisible connections. In other words, is the intuition of the random mixture primary or does the concept of chance have a history? And if so, what is its history? This is what we mean to establish by experiment.

1. *Technique of the experiment and general results.*

The child is given a rectangular box which rests on a transversal pivot, allowing it to seesaw. In a state of rest, the box is inclined to one or the other of its shorter sides, and along this width are arranged eight red balls and eight white balls, each group separated

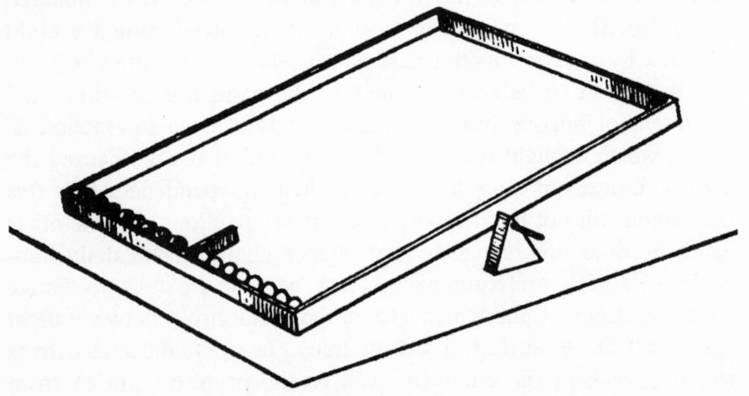

Figure 1.

by a divider (Figure 1). At each seesaw movement, the balls will roll to the opposite side and then will return to the original side when the box is tipped back, but in a series of possible permutations. The successive movements of the box ought not to be done too brusquely; in this way the mixture will proceed in gradual steps. For example, in the beginning, two or three of the red balls will be mixed with the white ones, and vice versa, and then the mixture will, little by little, be greater.

Before the box is first tipped (but telling the child or showing him how it moves while holding the balls in place), the child is asked a question: What will be the arrangement of the balls when they return to their starting places? Will the red ones stay on one side and the white ones on the other? Or will they get mixed up, and in approximately what proportion? We then proceed to tip the box and have the child note that two or three of the balls are in different positions. Then he is asked to predict the result of a second move which we then make, and so on. After some tries, he is asked to predict the result of a large number of moves of the box, and we note especially if he expects a progressive random mixture or a general crisscrossing of the red balls to the side of the white ones, and vice versa, and finally a return to their original order (that is, a final reordering).

Making a drawing of the arrangement of the balls can help in the questioning: He can draw his prediction of the first tipping of the box, and then after the first trial, make predictions by drawing the outcomes of successive trials. We will ask him in particular to draw an arrangement of the balls which he thinks is the best possible mixture. There is not always a connection between the drawing of the balls in their mixed positions and the trajectories of the balls which he is asked to draw. But an examination of this lack of agreement is quite useful for an interpretation of the thought process of the subject.

One notices that the different questions lead quite naturally to examining also the manner in which the subject conceives of the operation of permutations. We will return to this point in Chapter VIII, indicating there the technique to be used for the study of the development of these operations. We will note here only that it is a help to question the same children simultaneously by means of the tipping box of balls and also by means of the experiment with the permutations of counters (Chapter VIII). The correlation between the evolution of ideas concerning mixture and the development of the operations of permutation is an instructive factor in the interpretation of responses.

To remain for the moment with the mixture of the balls, the reactions observed at the time of the preceding questions permit us to distinguish three different stages. During the first stage (up to seven years), the mixture is conceived of as a total displacement of the elements, but without any intuition of a permutation of the individual positions nor any anticipation of an interaction in the trajectories. This total displacement certainly yields a state of disorder, but for the child it is not final and he often predicts that the balls will ultimately return to their original order. There is not, in the strictest sense, either a real mixture or chance. In the course of the second stage (from seven to eleven years as median), there is a progressive individualization of the positions, then of the trajectories, with the gradual construction of an intuitive scheme of permutations, but without a complete generalization. In the course of the third stage (over eleven to twelve years), the mixture is conceived of as a system of permutations due to the fortuitous collisions in the trajectories.

2. *The first stage (four to seven years): Failure to understand the random nature of the mixture.*

The reactions of the first stage are characterized by a very significant conflict between the facts noticed by the child which force him to see a progressive disorder and the interpretations which he searches for and which remain foreign to the idea of random mixture. In other words, the subject is obliged to accept the evidence of the changes of position due to the mixing, but he refuses to see in this a fortuitous mixing and decides to look for uniformities contrary to chance. Thus, in spite of the first permutations which the subject foresees, he predicts the return of the balls to their original places or perhaps a regular change of position, but not haphazard arrangements. For example, each ball is supposed to follow only a single trajectory—in particular, the red balls are to replace the white ones, and vice versa (Figure 2). A large number of movements of the box will not increase the mixture, and frequently the subject even expects a return of the balls to their original places. As for the drawing of the trajectories, rather than giving the child the idea of permutations, on the contrary, it leads him to the idea that the ball will come back to its starting place even when the fact of mixing has been previously established.

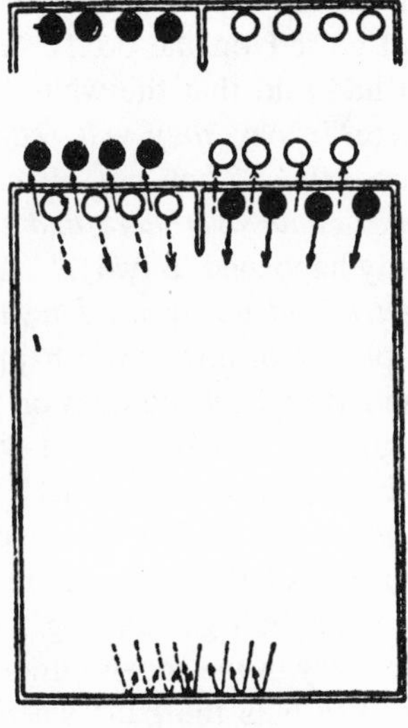

Figure 2.

Here are a few examples:

ELI (4; 4) begins by predicting that the balls will return to their places. *"They will go there* (a few centimeters from the point of rest)." And then: *'They will all come back in place* (perfectly arranged)." "Watch (the box is tipped and one white ball passes into the area of the red ones)." *"See, it's just as I said."* "Look again." *"One white ball went over there* (to the red side), *but all the red ones are there."* "And when I tip it again?" *"The white ones will be here and the red ones there* (complete crossing of all the balls)." "Take a look (another tipping and the mixture is greater)." *"Ah, it's all mixed now."* "And if I keep it up, will it be more or less mixed?" *"No. Don't do that* (he puts the balls back in place as if the mixture had bothered him)." "Now draw them for me as mixed up as they can be (he makes a drawing showing six white balls on one side, and on the other side three white ones together and three red ones)."

FER (5; 3): "What if I tip the box?" *"The balls will get all mixed up."* "How?" *"The white will go there and the red there* (crossing of reds to one side and white to the other)." (He makes a drawing of six reds on the right and six white ones on the left.) "Where will that ball go if I tip the box?" *"There* (return to the same place)." "You had said that the white balls would go there and the red ones here?" *"No, they will return to their places."* "Watch (the experiment)." *"They got mixed up. They crossed over and the red ones came over here and the white ones went there* (in fact, this only happened to two)." "And if we tip again?" *"They're going to get mixed up again. They'll go there and then there."* "Draw me a picture of how it will happen (he draws again six white balls on the left and six red ones on the right)." "Where will that white ball go?" (He draws a line which takes it to the other side on the right.) "And this red one?" *"It will go over and stop there* (inverse movement)." Fer continues the same drawing for the balls as follows: All the red ones move now to the left and all the white ones to the right making a complete crossover. At first there is a symmetry in the courses drawn, but after drawing some awkward lines, Fer puts the balls wherever he finds a free space which gives the appearance of there having been collisions in the trajectories, but the drawing is not done with the intention of showing any collisions.

VEI (5; 6): "If I tip the box, how will the balls then be lined up?" *"Just as they are now."* "Take a look (we tip the box: one red passes into the white section and one white into the red." *"They can't come back the same."* "And if we continue?" *"Then they will get even more mixed up."* "Why?" *"Because two of them will roll to the other side."* "And if I tip the box again?" *"Another one will move* (three red balls in with the whites and vice versa)." "And the next move?" *"Again one will move. They will get mixed even more* (crossing over) *and afterward they will start to come back to where they started* (by crossing in the opposite direction: the red ones will come back to their original position and the white ones too.)" "Watch (and we tip it again but this time there is only one red mixed with the whites and one white with the red ones)." "And if we continue?" *"It will stay as it is* (Vei holds with what he sees)." "And if we continue doing this all afternoon, what will it be like?" *"It will all come back in place as it should."* Again we ask him for a drawing showing the maximum mixture: All the red balls are shown on the side which originally held the white ones and vice versa.

MON (5; 8) foresees correctly, from the start, that the balls will not *"remain in their places"* nor will they be completely mixed. *"They will roll."* "Watch (the experiment: one red ball rolls in with the white ones and vice versa)." *"They didn't stay in their places; a red one is here and a white there."* "And if we continue, can we know what will happen?" *"No, no one can be sure because we aren't balls."* "Correct. But if we do it again, which will most certainly go to the other side, this red one which is in with the whites, or one of these whites?" (Hesitation.) *"The red one, because it has to be in its place."* "Do you want to try it?" (He does the experiment and one more red goes over with the whites.) "Now what do you say?" *"They know where to roll because they roll all by themselves."*

SCHU (5; 6): The same initial carefulness and the same tendency to believe that the balls will come back into place. He draws a series of trajectories going and returning: "Why? Will each one return to its place?" *"They know where to go because they're the ones who have to do it."*

CAR (5; 8): *"They're going to roll to the other side and then come back."* "How?" *"The reds on one side and all the whites on the other side."* "There is no chance that they will get mixed up?" *"Maybe not completely; two reds will go here and two whites there."* (We do the experiment: the mixture starts.) "And if we continue?" *"They'll come back as they were before."* "Watch (another move of the box: the mixture increases)." "And if we continue?" *"Still more mixed."* "And finally?" *"All the red ones will be here and the white ones there* (complete reversal)." "And then?" *"Like that* (a return to the original position)." *"It will come back to the way it was."*

TAI (5; 8) after having seen the mix start: *"It will get mixed up less as we continue."*

YVE (6; 2): *"The red ones will come back there and the white ones there* (as they had been)." "Will they come back as they were, or will they get mixed?" *"Come back."* "Watch (the experiment: one of each goes to the other side)." *"Ah! no, one here and one there."* "And if we continue, will they mix more or less?" *"More, I think . . . no, it'll remain the same."* "Watch (we tip the box)." *"It's even more mixed up."* "And if we continue?" *"It'll get more mixed up."* (The experiment: three reds on the white side and vice versa.) "And more?" *"No, it can't get more mixed up because it's already very mixed."* "And again?" *"They'll come back into place* (as before) *into the same slot."*

ROL (6; 3): *"They'll stay* (in their places)." (Experiment.) *"Just a few of the reds and whites on one side and the same on the other side."* "And if I continue?" *"All the red here and all the whites there* (crossover)." "They will separate then?" *"Yes, the red ones here and the others there."* (Experiment.) "Is that it?" *"No."* "And on the next try?" *"They'll come back in place."* The drawing of the trajectories shows a complete and quite symmetrical crossover.

These reactions present an undeniable homogeneity, in spite of individual variations, numerous samples of which we have been careful to give.

Taking the initial prediction first, we notice that the majority of subjects expect, before the experiment, a pure and simple return of the balls to their original slots. It is true that a certain number of subjects, from the ages of four to five, anticipate a mixing, but unwillingly, and they still have the feeling that, in spite of everything, a ball *"ought to be in its place"* (as Mon said). Finally, whether the child is reluctant to predict a mixing of the balls or whether he does foresee it to some extent, in both cases he does keep the idea of a connection either between the balls of a certain color and their initial slots or between the balls of the same color when they go over to the other side.

During the stages of the mixing, that is, in the predictions that the subject makes after the first tipping of the box and then after each of the following movements, it is remarkable that the child does not expect a progressively greater mixing. Rather, he thinks that (1) either the original distribution will be maintained (or will be reproduced), (2) he believes in a return to the original arrangement, or (3) he even minimizes the mixture by imposing on it a regular and in some way ordered process, that is, a general crisscrossing of the whites and the reds, or the displacement of one more element each time, and so forth. The subject Yve reflects the first attitude when he says that *"it will remain the same"* and especially when he says *"it can't get more mixed up because it is already very mixed."* The subject Tai is of the second type when he foresees, after the mix has begun that *"it will get less mixed as we continue."* The third reaction is the most frequent, for Eli and Fer (*"the white ones will go there and the red ones here"*) and even Rol who, after the same prediction in the same terms, is so surprised to see that the balls do not separate according to color that he cries, *"They will never get separated! No, never!"* Notice Vei also, who says that there will be a regular displacement of one more element each time the box is tipped until finally *"they will all go back to where they belong";* that is, the general crisscrossing will be reversed until there is a return to the original state.

We see then that the progressive mixing of the balls is either denied or thought of as too regular, which is simply another way of avoiding the idea of chance. Mixing is either an accident or a momentary disorder in which the elements themselves tend to take over (see the animism of Mon and of Schu), or even a transitional phase involving a general crisscrossing, and finally a crossover in the other direction to come back to the original state. Nothing is more curious than this last belief, so frequent and often so explicit: to continue the mixing will bring a *"straightening out"* as Car and others say. In fact, at this age level—four to seven—the child is still incapable of operative thought because he is unable to conceive of reversible operations. He believes, for example, that a shape whose form can be changed no longer has the same number of elements after the change because he does not yet understand that transformations in one direction can be reversed. It is, however, at just this age that he admits that the elements can return to their original places after a complete mixing, as if unmixing were the inverse operation of mixing. How are we to explain then this apparent contradiction?

In reality, the failure to understand the irreversibility of random mixing depends on the same reasons as does the failure to understand the reversibility of operations: In both cases the child makes a judgment about the perceived transformation without placing it into the whole system of possible transformations. In the case of operative irreversibility, for example, when six counters are first placed close together and then are spaced out, making the child think that there are now more of them, the transformations make different configurations, including the initial state. But the child considers each configuration alone, independently of the transformations which shaped it; that is, he sees it only as replacing the preceding configurations. For this reason the spaced counters are no longer equivalent to the tightly grouped counters and represent a new group, larger in number and substituted for the group from which it derived. It is the same situation in the case of the random mixture where there was a failure to understand the possible transformations. But it had an opposite result when considered in the light of the hierarchy of successive states because it is no longer the derived state which destroys the initial state, but actually the initial state which continues to take precedence over the next states. In fact, in the eyes of the child, the mixed state does not constitute a true state comparable to the initial order. Order remains privileged and mixture or disorder is a momentary alteration ("against nature" as Aristotle would say); the successive positions of the balls are not thought of as homogeneous results of comparable transformations. The reason for this interpretation by the child is precisely that he does not consider these transformations as the common source of the initial order and of the diverse states of the mix, that is, as actual operations consisting of permutations. If he envisaged them as such, he would understand that every change of order, or permutation, does, in fact, constitute a reversible operation, but without any priviliged state; and thus the different distributions of the balls would all be comparable because they constitute all of the equally possible different orders. But he would also understand that a mixture is irreversible in so far as it is simply one accidental arrangement among all the conceivable permutations and this random process leads to the most probable permutations without excluding the others. We can say then, and not as a paradox that, because the child lacks the operative understanding of permutations, he does not think of the mixture as irreversible ;and also that it is for this same reason (as we have seen elsewhere [1]) that he does not recognize the simple operations of placement or order as already themselves reversible. If, in fact, he were able to grasp the operations of order, he would anticipate the great number of ordered sequences that can be constructed with eight white balls and eight red ones and he would grasp then that a mixture is irreversible because it is tied not to a priviliged order but to the conditions of probability. But because the child reasons according to a hierarchy of states and not of possible permutations, that is, probability, he considers the mixture as due to the accidental displacements of some elements which tend then, by a sort of internal force or need, *"to come back to their places"* (Mon) or *"be as they should"* (Vei). This quasi-animistic intuition, therefore, has nothing to do with operative reversibility.

In brief, the apparent reversibility which the subjects of this age attribute to the random mixture constitutes in fact the contrary of operative reversibility; and it is only with an understanding of this process in the form of a reversibility of the permutations themselves, that the subject will finally grasp the probabilisitic reasons for the irreversibility of the mix.

Nothing is more helpful in this regard than an examination of the drawings representing either the most mixed state of the balls or the individual trajectories of the balls leading to the state of greatest mixture. Note first that at this level there is no necessary accord between these two drawings because the child does not often visualize clearly by what paths the balls will go from their initial orderly state to the state of maximum mix. This in itself is instructive and shows the nonoperative character of the transformation, that is, the failure to understand the idea of permutation. In the second place, at this stage, the drawing of the paths gives

[1] Jean Piaget and Bärbel Inhelder, *The Child's Conception of Space* (New York: Norton, 1967), chap. III.

only two possible models: Either the balls move out of their respective positions in a straight line and come right back in the same way (and this can be observed at times even in subjects who admitted a moment before the possibility of a mixture), or all the balls on the left move to the right and vice versa (for example, Fer). In this last case, the child proceeds to draw symmetrical and quite regular paths as if nothing was going to hinder the balls in their moves; they hit neither each other nor the walls of the box. These two kinds of drawings confirm quite well the absence of combinatoric notions and the failure to understand possible permutations as well as the failure to understand the idea of randomness. As for the drawings representing the maximum mixture, they are equally revealing: They always show a well-ordered distribution simply differing from the original order. For example, six white balls on one side and three white ones next to three red ones on the other (Eli); or a complete crisscross (Fer), and so on.

Perhaps it will be said that these last drawings show simply the lack of flexibility in operative representation and imagination at this age level, and do not exclude the understanding of physical mobility which the subject noted during the experiment itself. But if we compare these drawings with the totality of the reactions at this age, the over-all picture shows precisely that the lack of an internal mobility of thought goes along with the failure to understand the mobility inherent in the physical process of random mixture itself.

3. *The second stage (seven to eleven years): Beginning of the idea of combinatorics.*[1]

It is at about seven years of age that the intuition of chance appears along with the establishment of the first concrete operations, interrelated and reversible. This is confirmed by an increasingly correct prediction of the progressive mixing. But the elementary operations of order (to establish an order of progression ABC . . . , and to invert it as CBA) are far from suffi-

[1] The mathematics of permutations and combinations (translators' note).

cient for constituting a complete scheme of what are properly called permutations (as we will see in Chapter VIII). Thus, lacking such a scheme, the details of this gradual mixture which the child foresees are not yet sufficiently analyzed. In particular, the drawings do not always give individualized representations of the paths of the balls. There is still either a general crisscross (especially at the beginning of the stage, in substage II A), or collisions, but without a precise correspondence between the paths and the final positions considered characteristic of the best mixture possible.

Here are some examples of a first group of subjects (substage II A), who after predicting a mixture, end up sometimes allowing either a crisscross or a general return to the point of departure:

CHA (6; 6): We first give him six balls (three reds on the right and three white ones on the left): *"They'll get mixed up because they can't stay the same. They'll move around as they go across."* "In what way do you think will they get mixed up?" *"I don't know exactly."* (We do the experiment: A red ball passes to the right and a white one to the right.) *"It's mixed up. I said so."* "And if we do it again?" *It'll get mixed up again."* "More?" *"Don't know."* (The experiment ends up with two red balls next to each other on the left and two white ones not next to each other on the right.) *"Oh, yes."* "Could it be even more mixed up?" *"No, that's the best."* "If we do it again?" *"It will change because they can't stay in the same places."* "Do they know where they have to go?" *"Oh, no."* "And you?" *"No."* "And me?" *"Maybe. You're bigger than I am. You might know."* "And if we put in more balls, will it be better mixed?" *"Yes, more mixed up because the box would be less empty."* "O.K. Here are six white balls and six red ones. Arrange them as they would be if they were mixed as well as possible." *"That would be if we had the same thing on each side."* (He puts three red ones next to each other and three white balls together on each side.) "And if we tip it now several times, will it be better mixed?" *"Maybe, or maybe less well mixed."* "If we tip it many times, will they get separated?" *"No. Maybe once that could happen, but not many times. If we tip it many times, there is less chance."* The drawing of the best possible mixture shows finally two reds alternating with two whites, etc., on both sides, but without Cha coming to an understanding of how the trajectories could bring this about.

FRAN (7; 8): "How will they come back?" *"That depends. Some will come back to their places and some won't."* "How so?" *"They'll get mixed up. Perhaps it'll be that a red one will take the place of a white one, or even two."* (Experiment.) "What happened?" *"They aren't in their places."* "And if we continue?" *"Maybe they'll go back to their places, or maybe they'll get mixed up."* "What is most likely?" *"That they'll get mixed up. It is less certain that they'll go back into place."* "Why is that less certain?" *"Because while they are going to the other side, they can roll out of place. It's not easy for them to roll straight."* "And if I do it a third time?" *"They will get even more mixed up."* "And the fourth time?" *"One or the other; they'll get more mixed."* "And the fifth time?" *"Maybe more or less, or the same. Later they can get unmixed."* "And if I go on?" *"They will almost all go back to their places."* A drawing of the paths leading to the maximum mixture: All the red balls go over to the side of the whites and vice versa.

DUM (7; 11) starts with the prediction. *"All the red ones will go over to the other side and the white ones to this side."* "And afterward, at the next move?" *"Maybe they'll go back to where they were."* But after the first experiment, he predicts a progressive mixture. We ask him then to draw the maximum mixture; he draws a red alternating with a white, etc., on each side. But in the drawing of the paths leading to this alternating pattern, he shows a general crisscrossing: "This will give that (alternating pattern)?" *"Yes. As no."* He then thinks up a compromise between the two models.

AL (8; 2) predicts that *"It'll get a little mixed up."* "In what way?" *"One red on the right and a white on the left."* "Why?" *"Because the same thing has to happen on each side."* (We do the experiment and what he predicted happens.) "Were you sure about it before we tried?" *"No."* "And was I?" *"No."* "Can we know in advance?" *"No. It could get mixed up in a different way another time."* "And if we do it again?" *"They'll get mixed up more. They can roll where they want to."* "Could they get mixed up less?" *"Yes."* "Which is more likely?" *"That they get mixed up more.*" "Why?" *"Because they already got mixed up before. It's not certain, but I think so."* "And what if we continue to mix them up for a long time?" *"Finally they will all go back into place."* "Why?" *"It gets mixed up in every which way. Afterward it goes back together."* "Why? . . . What if we mixed up fifty white balls with fifty red ones for quite a while. At the end would they come out separated?" *"Yes."* "And one hundred white with one hundred red, would they also separate?" *"Yes."* "Why?" *"Because if this one can come back together, others can too."* "And if we had fifty red ones and fifty white ones, or eight of each, which would straighten out quickest? *"The one with the most."* The drawing of the best mixture possible shows an exact alternation of colors.

Bru (9; 3): "How will they come back?" *"It'll get mixed up."* "In what way?" *"A red one in the middle of the white ones, a white in the middle of the reds."* "Is that certain?" *"No, because that's chance. We never know how it's going to come out."* "Is there more of a chance that this will get mixed up, or not?" *"That they get mixed."* (Experiment: two red ones with the whites and three whites among the reds.) "And now will it get more or less mixed up?" *"A little more."* "Why?" *"It's already mixed up. Up there* (at the other side of the box) *it gets mixed up a little more."* "Draw it the most mixed up (he mixes the colors with no pattern)." "Is that the most mixed up possible?" *"No."* (He draws an exact alternation of colors.) "If I start again now?" (We move the box again which increases the mixture.) *"It will get less mixed because there are some reds which will go there probably* (into the whites), *and some whites probably there* (into the reds)." "Why?" *"They'd have to go there because there isn't much space between the balls* (thus a crisscross is seen as less of a mixture resulting from a continuation of the mixture itself). "And if we continue longer?" *"It will come back because bit by bit it will come back to the way it was."* "How can it get unmixed?" *"Because some going over there will make a few more over here* (shows the reds and whites each going back to their original places after the assumed crisscross)."

At (9; 10) admits as did the earlier subjects that there will be a progressive mixing, *"because they are already a little mixed and can get mixed up even more.*" "And if I continue?" *"They will get still more mixed."* "And if I continue until evening?" *"Still more and more. They will be all mixed up."* "Draw me now the paths they will take to give the best mixture possible (he draws quite a symmetrical crisscross)." "That's what you call the best possible mixture?" *'Yes, ah, no."* "Then draw me the best possible mixture (he draws a new crisscross)." "But that's the same thing." *"No, something else, they will be more mixed."* "Make a drawing to show where the balls will be following those paths (he draws an approximately alternating pattern)." "Draw for me now the paths which the balls will follow to get there (he again makes crisscrossing trajectories)." "If that's the path the balls will follow, will they finally come out as you drew (alternating pattern)?" *"Yes."* "Really?" *"Oh, no, the reds will go over into the white spaces and the whites here where the reds were."* "O.K. Start over again. You've got to draw paths which will take the balls to where you showed them in your drawing (he again draws crisscross trajectories)." "Are you sure that this will give the design that you drew?" *"Not exactly."*

A second group of children of a slightly higher age level (substage II B) have moved away from the idea of a general return as well as from drawings of almost exclusively crisscross trajectories and try to show how the balls collide with one another. But they are not yet able to show an exact correspondence between the trajectories and the final places in which the balls are shown:

RIC (8; 5) foresees a progressive mixture. "Will it get more mixed up if we tip it ten or twenty times?" *"Twenty times because there will be more which get out of place and go over to the other side."* But he finds great difficulties in drawing the trajectories: *"Oh, this is not simple. I have to do a lot of thinking."* He does, however, start to show collisions in the trajectories, but is not satisfied with it. He draws next the positions of the balls and ends up with a crisscross before coming to an irregular alternation of colors. As for a complete return to their initial positions, that is possible, but only as a kind of exception: *"Certainly, but we would have to tip the box a number of times because otherwise it would be very hard for that to happen."*

MAR (9; 11): *"They'll get a little bit mixed up."* (Experiment.) "Yes, and after more?" *"They will be all mixed."* "In what way?" *"Three reds with the whites, five whites with the reds."* "And if I keep it up?" *"Maybe they will be even more mixed."* "Why?" *"They won't find any space, they'll hit each other and they'll go to another place."* "The fourth time?" *"Even more mixed."* "Why?" *"They won't find any space, they'll hit each other and they'll go to another place."* "And the fifth?" *"Maybe they'll be completely mixed and all the reds will go there and all the whites here* (crisscross)." "And if we continue for a long time?" *"They'll maybe come back to where they started."* "After how many times?" *"I don't know; maybe fifty times."* "What's most probable?" *"All changed, all in their places, you can't say. Most probably they won't be in their places."* Now the drawings represent an irregular alternation of colors, and the trajectories show a mixture of whites and reds: *"I should have done something else. This way they won't get mixed up. I couldn't get them to come back like I wanted: The white ones came back to their country and the red ones to theirs."* "O.K., draw them all mixed up." *"A white one takes the place of the red,* etc. (it ends up with something like a general crisscross)." "Like that?" *"A white one takes the place of the red,* etc. (it ends up with something like a general crisscross)." "Like that?" *"No, on another picture we could have made some white ones go into the reds, some reds which go in with the whites, and some white ones into the other part."* This last plan is correct, but is thought of only after two wrong systems.

These then are the principal reactions of this second group classified according to the two levels II A and II B. Compared with those of the first stage, the main characteristic of these subjects is to believe in a real and progressive mixture and not only an apparent and regressive one. Moreover, from age level II B, the drawing of the trajectories shows an individualization of the paths taken, shows collisions, and is no longer simply a global displacement of all the balls of one color with no collisions. In other words, from age level II A there is the beginning of the idea of randomness with a progressive refinement of the idea in the course of substage II B; stage II marks, therefore, the formation of the idea of physical chance in a first and elementary form of mixtures. At level II A it is true that the child is still very attached to the idea of a general crossover and often even to the idea of a complete return of the balls to their starting points. These attitudes are a prolonging of the first stage and they are seen as well in the drawings of the trajectories which do not show individual collisions. But instead of conceiving the general crossover or return to the initial point as excluding the progressive mixture or correcting or directing it, these interpretations from now on are given as probable or final end products of the mixture itself. This is a more probable result at the beginning of stage II A, and less and less so in later ones. At stage II B, the return to the initial state is thought of as a simple, if exceptional fluctuation.

First we must insist on the understanding of mixture itself as shown in the remarks of Cha, Fran, Dum, Al, and others. The two decisive remarks in this regard are made by Al who insists on the multiplicity of possible arrangements of the mixture (*"It can still get mixed up in a completely different way"*); and Bru (much like Al) who underscores the impossibility of predicting with any certainty (*"because it's chance, and nobody ever knows how that will turn out"*). This is quite different from the subjects of stage I for whom order and disorder do not exist (the latter constituting a momentary alteration of the former). The children of stage II A accept randomness as a positive part of reality because they are able to foresee a large number of possible distributions in the movements of the balls and consider all these possibilities as being equal or equivalent without any special meaning for the initial state; hence, the impossibility of foreseeing the details of the transformations, which are due to chance (note that the word appears at about eight to nine years of age, while the concept was not understood in stage I). But for these subjects, if randomness excludes the prediction of each change in the arrangement, it does, however, call for a judgment of the total probabilities concerning the increasing nature of the mixture. Al formulates explicitly the principle of progressive mixture: *"because they're already a little mixed up. . . . In a while they'll get even more mixed"* and even *"all the time more."* At level II B it is even more obviously so.

Now, although the child at level II A may well predict the fact of the progressive randomness of the mixture, he will conclude, nonetheless, that the continuing mixing of the balls will end up with a general crossover and finally the balls will come back to the initial order. The differences in the situations show quite well that this is no longer the same belief that we saw in stage I and that a nuance has been introduced: While in stage I the crossover and the return showed a tendency to order contrary to the accident of the mixing, in stage II A it is these same causes which control the mixture and bring it eventually to a general crossover and then to the original positions. This is clearly stated by Cha, Fran, Al (*"It gets mixed up in all sorts of ways; and then it comes back together"*), and especially Bru: *"little by little it'll slowly get unmixed up . . . because going over there* (opposite side to the original position) *there'll always be a few more there* (inverse of the inverse equals the original)."

We see here the limits of the progress accomplished in substage II A. At this stage the mixture is accepted as a positive fact and is no longer, as in stage I, a kind of antinatural state, a momentary interruption in the spontaneous and obligatory return to order. But if it is admitted, it is not yet understood. To understand it, the child must assimilate it not only as a concrete operation which assures the reversibility of an elementary order (changing ABC to CBA, etc.), but as an operative scheme of permutations, as we will see in Chapter VIII (eleven to twelve years); the subject would have to grasp, in addition, the nondirected character of the permutations involved, that is, their fortuitous nature due to encounters and collisions occurring in the course of the trajectories and producing the interactions. Now, because he fails to realize these conditions, the subject still sees the mixture as resulting from the simple passage of balls from one side of the box to the other. The more mixed the mixture becomes, the more red balls will go to the side of the whites, and vice versa. It is obvious then that the mixture itself tends to a general crisscross, and no less obvious that the logical next step to this crisscrossing of the whole is to return to the point of departure, that is, to a new crisscross but in the inverse direction, thus canceling the first one. We refused to admit operative reversibility in stage I since the return that the children talked about was only a returning to the original state after accidental mixing; the general return spoken of by subjects of stage II A is, on the contrary, a good example of what we properly call reversibility, since it is actually a double crossing, that is, the inversion of a direct operation which leads to the identity operation. However, it is too simple an operation and this is where it is in error. The child reasons on simple displacements of the whole, whereas, to see it clearly, he would have to invoke multiple permutations between the elements themselves considered individually.

The reason for this error is equally clear. It comes from the same causes as those which kept the subject from realizing the system of permutations when he was asked about this kind of operation (see Chapter VIII), that is, from the difficulty of reasoning simultaneously on several displacements at the same time. In the case of the mixture, this difficulty is presented in its simplest and most crudely childish form: The child imagines the displacement of each ball as if the others stayed still during this movement. More precisely, he can begin to think about the simultaneous movements of the balls of one of the two categories (reds or whites), but he does not remember that the balls of the other color have already left when the first ones are in play and that they collide then in the course of the trajectory. This is why there is no individualization of the moves (an absence that we noted earlier as characteristic of this stage) or permutation between isolated elements. All the red balls move little by little among the white ones, as if nothing prevented this, and all the white ones move reciprocally among the red ones as if these latter were already in their places once and for all, or as if they were waiting before going into other places themselves. The trajectories are, therefore, not yet independent or interacting, but the balls proceed by categories (red or white), even though moving successively (a red and then another, etc.).

Finally, we should note that certain subjects at this II A level, when they are not deciding on the itineraries, but only on the final positions, are able to show sooner or later the maximum random mixture as something different from the simple crisscross. They see it then as a more or less regular or irregular alteration of colors. But when they try to explain how that happened, they fall back on earlier hypotheses.

This brings us to level II B which is a quite constant continuation of stage II A. What is new in this next stage is that the child begins to see the permutations in detail: The balls *"will start to get mixed up a bit,"* said Mar, *"they are coming up against one another and they are going to another place."* From then on the subject rejects the crossovers as inevitable and especially rejects the return to the initial position, although considering such a return as possible but very improbable (*"it's very difficult for that to happen"*—Ric). This notable progress in predicting is marked in particular in the drawings of the best mix possible. In general this drawing will show irregular alternation of colors. But an interesting thing, which indicates exactly the continuity with the preceding stage, is that the child at this level, II B, does not yet succeed in making the drawing of the trajectories coincide with the drawing of the positions. At times the first is ahead of the second; ordinarily it is the reverse: Mar gives a particularly clear example of these difficulties of representation.

This discordance between the drawing of the mixed positions and that of the paths confirms in the most precise manner what we said about the difficulty that the subjects of stage II have in conceiving several different and simultaneous systems of displacement. The mixture of paths supposes, in fact, more than a static intuition of the finished mixture; it implies a play of fortuitous permutations, that is, again the simultaneous representation of several distinct movements (since each ball follows its own path), and yet connected (since the collisions make them act on each other). Now the child may be quite aware of the fact of collisions and their effects on the mixing, but he will not succeed in incorporating them. He knows that there are permutations, and he draws the final positions as a function of the permutations, but he still draws the trajectories as if each ball moved according to a regular plan of the whole group (Figure 3).

The persistence up to the age of about ten to eleven years of this type of difficulty constitutes a first aspect of the opposition between the concrete operations and the formal ones. We will often call attention in this work to this opposition. To understand that the balls from the left do not get over to the right side before those on the right have moved, and vice versa (that is, to under-

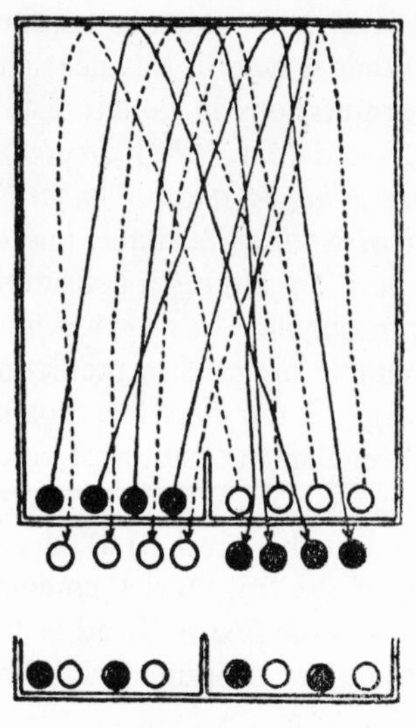

Figure 3.

stand the mutual dependency of the two motions), means that one has had to reason about two systems of reference which are, at the same time, both different and interdependent. And this is, no doubt, the proper character of formal reasoning. Herein lies a small fact but a revealing one, to the extent that it is closely related to what we will see next.

There is, therefore, nothing surprising about the fact that the children at stage II do not yet understand anything at all about the law of large numbers while they grasp quite well the role of number in small aggregates. We will, in fact, see that the law of large numbers also requires the use of formal thought. Our subject Cha agrees that, by adding balls to our apparatus, it *"will get more mixed up because the box will be less empty,"* but if we tip it several more times, it will be *"perhaps more mixed up or maybe less."* The return of the balls to their initial state seems possible after just a few motions of the box, but *"if we tip it several times, there is less chance of its happening."* In the same way Al guesses that the unmixing will be the same for fifty and one hundred balls as for a few *"because if these come back here, the same thing can happen with the others.*" There is even a better chance for a rapid separation with fifty than with eight because "*there are lots of them.*" With small numbers, however, the child understands quite well that the mixture increases with the number of moves of the box: According to Ric the mixture is better after twenty moves than after ten "*because there are more that get out of place and go over to the other side.*" Thus we have what could be called a law of small large numbers, but it cannot be generalized for true large numbers because of the lack of formal combinations.

4. *The third stage (from eleven to twelve years): Permutations and interaction of trajectories.*

Thus it is only at about the age of eleven or twelve, when formal thought first appears, that the process of random mixture is understood because, at that age, the observed facts are assimilated in an operative scheme based on the mechanics of permutations. First, here are a few examples:

STO (11; 1): "*They'll get a little mixed up.*" "And if I do it some more?" "*It'll get more mixed up.*" "And if I do it more?" "*It'll still get more mixed.*" "And if I do it for a long time?" "*Still more. If you have a lot of patience, it could come back together.*" "Is there much chance of that happening?" "*No, none.*" "But even one chance, or none at all?" "*One.*" "How many moves would it take?" "*A thousand.*" The drawing of the best mix shows an alternation of colors and the paths are completely intertwined.

ROS (12; 0): "*I'm sure that it is going to get mixed up.*" "And if I keep it up?" "*Still more.*" "Can they come back to their places?" "*Oh, not to the same places.*" "The red ones here and the white ones there?" "*Oh, no, they're going to hit each other.*" "And what if we keep it up until evening?" "*They'll just get more mixed up.*" Drawing: collision of trajectories, and a correspondence between this and the alternation of positions.

JEN (13; 6): Same reactions. "If we tip the box several times, do you believe that they'll return to their original places?" "*Oh, no, I'm sure they won't. That will never happen.*" "And if we do it a thousand times?" *"Maybe. In a thousand times that could happen."* "Ten times?" *"No. If we do it ten times, it won't go back the way it was."* "Are you sure or do you just think so?" *"I'm not sure, but it's hardly probable."*

VAL (13; 10): Same initial reactions. "Could the colors separate out?" *"No, I don't think so."* "What if I tip the box several times?" *"No."* "Fifty times?" *"No."* "And if there were only four balls of each color, could they separate out?" *"Yes, that could happen."* "And with five balls?" *"After a long time."* "With six?" *"I don't think so."* "If I tip it two thousand times?" *"Maybe, but it will be difficult."* "And if there are eight balls?" *"You'll have too many balls then. There'll certainly be some that go over to the other side."* "And if I tip it a thousand times?" *"It will be very difficult."* Drawing: zigzag trajectories, but collisions and interactions which correspond to the final alternating positions.

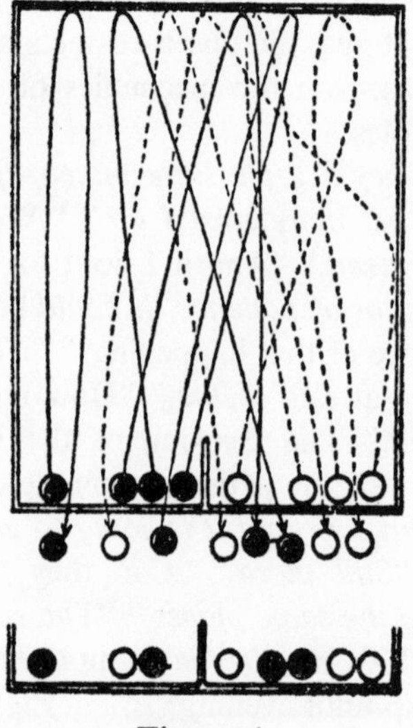

Figure 4.

We see the progress accomplished by stage II. The subject will understand from now on the collisions of the balls which he assimilates into a system of varied permutations, and his drawings now show an exact correspondence between the paths and the random mix of the final positions (Figure 4). And above all, he understands the law of large numbers: Everything is possible if we augment the number of moves (or reduce the number of balls), but the return to the point of departure is very improbable.[1] To the extent that we exclude the cases which are too rare and remain within the scale of usual approximations, the mixture becomes irreversible.

The discoveries which are part of stage III bring up the problem of understanding what mechanisms of thought are behind the discovery of the operation of permutations, and also of large numbers. We see at once, given the age of the appearance, that it is a question of formal mechanisms. But why does it happen this way? And what is the mode of action of these formal operations? In the following pages we will be able to be somewhat more definite about these things.

[1] We know that for ten red balls and ten white ones, we would need, on the average, 184,756 moves to get one chance of bringing the balls back to a state equivalent to the initial state, that is, with a separation of colors (whatever might be the internal order within each group of colors). The probability of return then, using twenty elements, is 1/184,756. See E. Guye, *Physico-Chemical Evolution* (New York: Dutton, 1925), p. 41. (It is this problem of ten white grains and ten black grains mixed to form a gray powder which gave us the idea of the present experiment.)

II. Centered Distributions (Normal Curves) and Uniform Distributions[1]

The problem of random mixture brings up the problem of the forms of distributions since the final positions of the elements in the mix, as well as the trajectories leading to these positions, necessarily take on certain distributive forms of the whole. For example, the grains of sand running out a small hole in a funnel or spread by hand over a road, will offer either a distribution centered around a hill (whose summit would match the opening in the funnel) or a uniform distribution as a function of the road's surface. If the forming of the idea of chance depends on an intuition of random mixture, as Chapter I has just allowed us to see, the fortuitous distribution which constitutes the mixture will bring up sooner or later in the mind of the subject questions of probability tied to the form of the whole in this distribution. Thus, to understand the mechanisms of a distribution centered in a pile, it is a question of seizing the symmetry of the possible paths of the grains of sand running out of a funnel. The whole takes the form of a regular pile because there is the same possibility for

[1] With the collaboration of Myriam van Remoortel.

each grain to fall either to the right or to the left, or to the front or to the rear. In the same way, the dispersion of grains of sand on a road, or of drops of rain on the surface of flat tiles, will be understood as being uniform when the subject recognizes the equal probability for each square meter or square decimeter of surface area to receive the same number of elements. It is these two problems of centered and uniform distributions whose solutions we would like to study here as a function of the mental development of the subjects.

The distribution in the form of a heap corresponds to the mathematical bell-shaped figure known as a normal, or Gaussian, curve. To get beginners to see the significance of this curve, we often use an instrument formed from a wooden inclined plane with nails in the form of a quincunx (five nails, one at each corner of a square and one nail in the middle). A certain number of balls are rolled down the plane, fed by a funnel whose opening is in the center of the upper part of the incline. The balls are dispersed at random and finally roll into narrow slots at the bottom of the incline. The center slots receive the majority of balls while the extremities receive the least; and the columns of balls form a bell-shaped distribution. The line running along the top of each column forms an empiric Gaussian curve, more or less regular.

We used an analogous but simplified machine to study our subjects' understanding of distributions centered in a pile. The advantage which such a method allows is to give an analysis of the symmetries which is essential for the solution of the problem, but using only two dimensions instead of the three dimensions as is more usual (pile of sand, etc.). The drawback in this mechanism is that the complex factor of the incline affects the balls. The incline is steepest between the funnel, which is above the middle of the lower part of the box, and the middle of the lower part of the box. But this physical factor which will be understood only later does not keep the general tenor of answers from expressing the notion of chance with a regularity quite comparable with the reactions described in the other chapters of this work.

As for uniform distribution, we were satisfied with a very simple mechanism which imitated the drops of water falling on square tiles: Small glass beads which do not roll too easily are dropped from a kind of trellis onto a sheet of paper divided into squares and the child is asked about the probability of uniform distribution as a function of the greater or smaller number of beads dropped.

The significance of the two problems studied in this chapter is then, if intuition of random mix is developed only very gradually, as we determined in Chapter I, will the subject succeed better in seeing the distributions of the whole either as a function of symmetrical displacements dispersed from a common center of origin or as a function of uniform dispersion? An analysis of these two questions will permit us, in particular, to push the study of the formation of the idea of chance in the direction of the role of large numbers.

1. *Centered distributions: Experimental technique and general results.*

We use five inclined boxes without lids which are shown to the subjects one at a time. The first four have a funnel-like opening in the middle of the upper part of the incline while the opening of the fifth box is placed on the right side (Figure 5).

The child is first shown box I whose lower side has a partition dividing it into two slots each the same size. We put a single ball into the funnel, and then a second one, and so on, asking each time which slot the ball will go into and why. Once we know these predictions and their motivations, we say that we are going to let them all go (about sixty small, round, metal balls of the same diameter) and we ask for a detailed explanation of the resulting distribution. We then do the experiment with the child and ask him to interpret the results.

Next we use box II (three slots of equal size, the one in the middle facing the funnel mouth), then box III (with four equal slots), repeating the same instructions. Then we show him a large box, number IV, provided with nails (as explained above) whose lower part has eighteen slots. With each box we examine how the subject uses the preceding observations. In addition, we

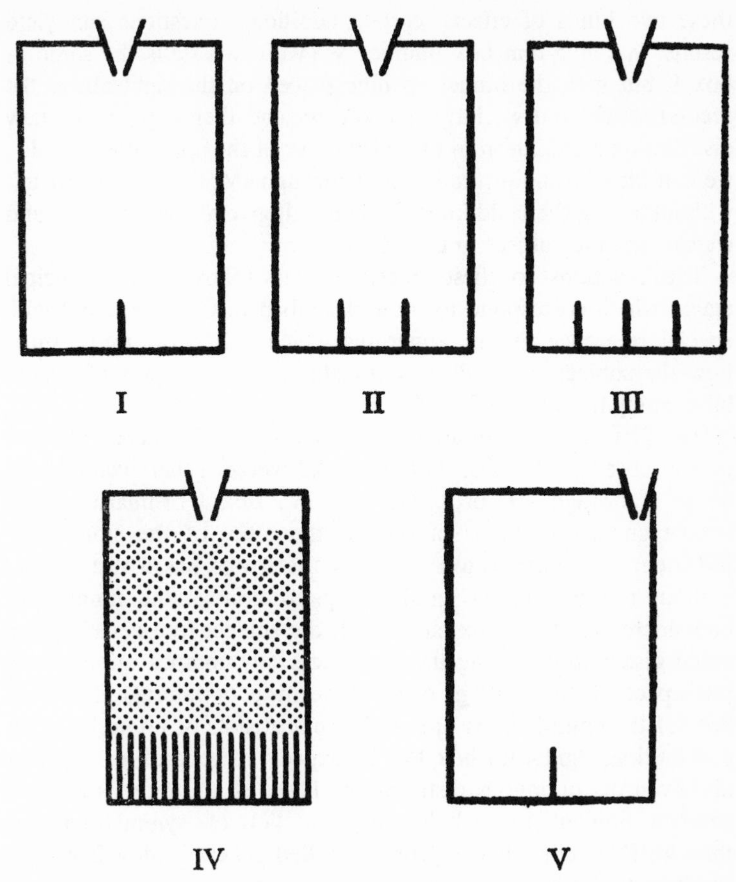

Figure 5.

are careful to ask the subjects after their explanations of each experiment if we would get the same arrangements of the balls regarding symmetry and the peak of the distribution, were the experiment repeated.

These different questions are sufficient when the child understands the idea of the existence of a maximum central frequency (the summit of the pile) and the existence of symmetry in the lateral slots. But, when it is a question of analyzing with him these two kinds of effects, certain additional questions can yield results. We give him box number V (with two stalls the same as box I, but with the funnel opening placed on the right above the second stall), and we have him predict and then explain the new distribution resulting from the asymmetry of the apparatus. Finally, we can tilt any one of the boxes, I through IV, to the right or left without letting the child know it. This will give a skewed curve and we can ask the subject to explain it.

The reactions to these questions will follow three principal stages which correspond to those described in Chapter I, but with special reference to this apparatus which will allow us to study how the subjects generalize or transfer their judgments of probability going from boxes I to IV.

The first stage (up to about seven years of age) is characterized by the absence of a distribution of the whole, generalizable from boxes I through III, or I, III, and IV. Box I ordinarily elicits simple guesses in favor of one or the other of the slots, or a judgment of compensation (such as "each one has a turn"), but without recognizing the gradual equalization as the number of balls increases. With box number II, the subject foresees either an equal distribution among the three slots (by compensation), or a pile-up of all the balls in one slot (central or lateral). But with box III the child bets on one of the central slots or on an irregular distribution. And with box IV, he expects in general an irregular and even discontinuous distribution. There is, therefore, neither a configuration of the whole, uniform (I) or symmetrical (II through IV), nor what is properly called generalization from one situation to the other.

Stage II (between seven and eleven years of age as median) is, on the contrary, characterized by the beginning of an understanding of the distribution of the whole which the subject generalizes. This is recognized in particular by the fact that the subject foresees an inequality between the central frequencies and the lateral ones. But this distribution remains insufficiently quantified because of the failure to understand the role of large numbers. For this reason, even if the subject sees a symmetry of the whole, he does not yet see any equivalence between the slots located symmetrically with respect to the center. In the presence of box I the subject ordinarily expects an approximate equality between the right slots and the left, but without any equalization yet which would be progressive with the number of balls. With box II, he certainly foresees a maximum in the central slot but not equality between the lateral slots. With box III, in the same way, there is an exact prediction about the privileged position of the two central slots, but without an equivalence between them, nor between the lateral slots. Nor does box IV allow for the prediction of a regular symmetrical distribution, but we begin to see here a transfer of the previous experiments on the configuration of the whole.

The third stage, finally (from eleven to twelve years of age), is marked by the quantification of the distribution of the whole, that is, by the prediction of an equivalence in the dispersion between the symmetrically located slots. This is clear for boxes I through III; box IV allows for some beginnings of gradations up to the discovery of the bell-shaped curve distribution. But the most remarkable progress is the understanding of the role of large numbers in the regularity of distributions, and this progress is related to the preceding quantification.

2. *The first stage: Absence of a distribution of the whole.*

The reaction at this stage is to be able to predict each isolated path without coordinating the whole and also to imagine the path followed by the whole group as if it were only one or just a few isolated elements. (This kind of reaction is not in contradiction with the failure to individualize trajectories, characteristic of stage I in Chapter I, since here the questioning bears first on each ball singly):

MAN (4; 10): Box I: "Where will the ball go?" *"Maybe there, or there* (he points to the two slots)." (Experiment: R.[1]) "And the next one?" *"There or there."* (Experiment: L.) "And the

[1] Abbreviations: R: slot on the right; L: slot on the left; M: single middle slot; MR: middle right; ML: middle left.

next one?" *"There or there."* (Experiment: R.) "And all the balls?" *"Don't know."* "Will we probably get more on one side or the other, or will they be the same?" *"Maybe the same. More will go there* (R)." (Experiment: same on each side.) "Why do we have the same on each side?" *"I don't know."* Box II: "Where will the ball fall?" *"There, or there, or there."* "And another?" (Same answer.) "And the whole box full?" *"All there* (M)." "Are you sure?" *"Or there* (L), *or there* (R)." "Watch." (Experiment: Left is the same as right and the middle one is greater than the side ones.) "Why is it like that? Why are there more in the middle? *"Because."* Box III: *"There, or there, or there."* Man does not choose then between the different possibilities and makes no generalizations about what he has noticed as he goes from one box to the other.

EIS (5; 1): Box I. Single balls: *"There or there."* All the balls: *"There will be more balls on one side."* "Which side?" *"We don't know because we haven't sent any down yet."* "How will it be, the same on each side or not?" *"A lot more on one side."* "Is that possible?" *"Yes, because sometimes the balls will roll like that* (R)." "Why?" *"Because there are lots of balls."* (We send half of the number down.) "You see, it's the same on both sides. Are you surprised?" *"No, because we didn't drop all the balls."* "And if we send them all?" *"There'll be a few more on the right."* (Experiment.) "You see that it is the same. Why?" *"Because I was wrong. Because there are a lot of balls."* "Look at this box (box V). What do you see?" *"The opening is on one side."* "Where will the balls go?" *"Here* (to the right, the slot just under the funnel)." "Fine. Now then with box I why were there the same number of balls on each side?" *"Don't know."*

Box II: "What if we drop them all?" *"They will all go here* (M) *because it's the middle."* "Will there be any on the right or left?" *"None at all."* "Why?" *"Because they will go in the middle."* (Experiment.) "Was that right?" *"No, there are some on both sides."* "Why?" *"Because they hit each other."*

Box III: *"Most of them here* (MR), *fewer there* (ML), *still fewer there* (L), *and fewest of all there* (R)." Eis comes to an irregular figure, therefore, without generalizing from what he has just noticed about the symmetrical form that box II gave. There is, nonetheless, a transfer concerning the choice of a slot near the mouth of the funnel for the maximum number of balls as well as having some balls on both right and left sides. "Why will there be more here (MR)?" *"Because the balls will go like this* (he shows a path from the funnel mouth to MR)."

Box IV: He predicts the maximum number under the funnel, and an irregular distribution for the rest.

FER (5; 4): Box I: "If I let them all run down, where will they go?" *"There and there."* "The same number on both sides?" *"I don't think so. Not many there* (L), *and many there* (R)." "Why more on the right?" *"If they roll down, they will stop there* (the top of the partition between the two slots) *and fall there* (R). *And one will stop and roll there* (L)." (We do the experiment.) *"Ah! The same on both sides."* "What had you said it would be?" *"More on the right."* "Why is it the same on both sides?" *"Because they all went there, and there."* "And what will it be like if I start over again?" *"I don't know at all. They will roll down before we can say it."*

Box II: *"Many will go there* (M) *and many there* (L) *and many there* (R)." "The same in all three?" *"I don't think so. Yes, in all three."* (Thus he makes the transfer from the equality he noticed with box I.) (We do the experiment.) "Is that right?" *"No, many here* (M), *and not so many there* (L) *and there* (R)." "Why?" *"Because the hole is there, and then they went there* (M), *and the others hit the walls and went there* (L) *and there* (R)." "Why are there as many there (L) as there (R)?" *"Don't know."*

Box IV with eighteen slots: *"Many will go in the middle, many here* (L) *and many there* (R)." (Thus he keeps the initial prediction of equality from box II.) "The same in all?" *"No, I don't think so. Not so many here, here, here* (pointing to the even-numbered slots)." "Why?" *"The nails will send them there, there* (pointing to the odd-numbered ones)." "They won't send any here and here (the even-numbered ones)?" *"No."* "Why? . . ."

Box III: *"Many here* (MR), *not so many here* (R), *here all full* (ML), *and none at all there* (L)." (We do the experiment with box IV.) "You see, it's the same on both sides (regular bell curve). Now if I do the same thing with this one (box III), what will it look like?" *"Not a lot any place."* "Here (box IV) where are there the most?" *"in the middle."* "Then there (box III), where will there be the most?" *"No place."* "Will it be the same as box IV?" *"I sure don't know."*(We bring out box II again and show it to him.) "Where will they go?" *"To the middle."* "And with this one (III)?" *"Mostly here* (ML), *because the little hole is just above it, here* (L), *not many, here* (R) *even less, there* (MR) *more. Maybe there* (ML) *less and* (MR) *more because the hole is right across from it."*

MOR (5; 6): Box I (after tests with single balls): "And now all of them?" *"There or there. More here* (R)." (Experiment.) *"Ah! no. The same in both."*

Box II: *"More in the middle."* "Why?" *"Because."* "And on the two sides?" *"A few." "Same on both?" "No, a few more here* (R)." (Experiment.) *'No, the same!* (R and L)."

Box III: *"More here* (MR), *fewer there and there and there* (L, ML, and R)."

Box V (the opening over R): *"There or there."* "More where?" *"I don't know."*

Box IV: arbitrary answers.

GON (6; 3): Box I: "Where will the first ball go?" *"There or there."* (Experiment: R.) "And another one?" *"Here* (L) *because there's already one there* (R)." (Experiment: R.) "And another?" *Again on the right because I was wrong, and if I was wrong, that means that it will always go there* (R)." (Experiment: L.) "And another one?" *"To the left because there are already two* (in R), *not to the right."* (Experiment: L.) "And another one?" *"To the right, so that there'll be three there, and afterward, three there."* (Experiment: L.) "And if I let them all roll down?" *"I don't know, to the right, because . . . no, I don't know. All over."* "More on one side?" *"Yes, on the right; no, the same because the same ones are rolling down."* "And if I do it ten times?" *"Not the same. Yes."*

Box II: "If I let one ball roll down?" *"In the middle."* "And another?" *"To the right."* "And another?" *"To the left."* "And if I let them all go?" *"Everywhere."* "The same number in all the slots?" *"Yes, they'll slide all over and they'll go like this: more in the middle, fewer on the left and even fewer on the right."* (Experiment.) "Is it like you said?" *"Yes, more in the middle."* "Why?" *"Don't know."*

Box III: *"Here the most* (ML), *not there* (MR)." "More, or the same?" *"The same, because it's the middle."* "And the others?" *"More on the right than on the left."* "And if we start all over again, will it be the same or not?" *"Yes, because we will still have the same boxes."* (Experiment.) *"Ah! No."*

Box IV (eighteen slots): "Where will the balls go?" *"Everywhere down here."* "Where will there be the most?" *"Here* (8, 9, 10)." "And the fewest?" *"There* (15, 16)." When we let the balls fall with the box slightly tipped to the right, giving a very clear asymmetry, Gon sees nothing abnormal about it. On the contrary, he sees, as with box V, that the balls have to fall *"more to the right because that is under the aperture."*

Such are the characteristics of the stage from four to seven years of age. They are most interesting from the point of view of considering the pattern of the total number of balls as well as from the consideration of the action of each experiment on the next one.

From a consideration of the prediction of the single trajectories, on the contrary, we notice only some attitudes whose results we will learn more about from later chapters (especially Chapter III). The subjects' reasoning is determined by two opposite explanations for their predictions: repetition or compensation. In fact, in several cases the child concludes from the ball having gone into one of the slots that following balls will have a tendency to go there also, while in other cases he draws the conclusion that they will go to the other side to balance the situation. For example, after Gon has seen a ball fall to the right in box I, he concludes that the following one will go to the left *"because there is already one there."* But since the second ball falls to the right also, the child shouts that he was wrong and that *"this means that it will always go there."* We have in this case, therefore, both a prediction by compensation and then one by repetition. We will try to analyze in Chapter III the motives for such reactions.

But in the case of letting all the balls roll down together, the reactions at this stage are quite significant and show conclusively a failure to predict and a failure to generalize after having noted the distribution of the whole. We must remember that box I with two slots has a single probability, L = R (that is, half for the R slot and half for the L slot). Box II with three slots has probabilities of M > L and M > R and L = R. Box III, finally, with four slots has three probabilities: ML = MR, L = R and ML (or MR) > L and > R. Therefore, only box I gives a uniform distribution, while boxes II and III (and IV) give bell curves centered around a maximum frequency in the middle. Now, whether it is a question of box I or of boxes II to IV, the common trait in the observed reactions is that the child at this stage does not come to a prediction of any regular distribution of the whole. In the case of the equality of box I, he ordinarily foresees an inequality of balls in the two slots; and, in the cases of the inequalities of boxes II through IV, he predicts either an equality or nonmotivated and asymmetrical inequalities (leaving aside for the moment the influence of each observed experiment on subsequent predictions).

In the case of box I, in fact, each one of the subjects we have reported (except Gon who hesitated) believes in an inequality of distribution between left and right and tries to guess which will be the privileged side, without thinking of symmetry. As for box II, the subjects say either that all the balls will be in the middle (for example, Eis, or Man before he corrects himself) or that there will be the same number of balls in the three slots (Fer), or even an inequality in the three slots, but without symmetry in the outside slots. Box III, in a similar way, brings out the following predictions: all in one of the slots (Man), maximum in just one of the central slots, and asymmetrical in the outside slots also (Eis and Mor), same reaction but with one slot empty (Fer), and finally, with Gon (who overlaps the following stage), equality for the middle slots and inequality for the outside ones. Box IV elicits responses which show all forms of distributions, but irregular, asymmetrical, and even discontinuous; however, these responses have the tendency of predicting a maximum in the central slots.

The first question to be asked on these predictions is that of the maximum. There is no doubt that from the initial stage the child tends to take into account the position of the mouth of the funnel. Box V, which has the mouth to the right, elicits almost always relatively correct answers. As for boxes II to IV which have both more than two slots and the funnel in the middle, the most numerous predictions are for a maximum in the central area. This is a quite natural situation, however, because without understanding the role of the slope (which is never mentioned at this stage), the subject sees no doubt that the path leading from the funnel to the middle slots is the shortest one (compared with the outside ones): For example, Eis, in talking about box III, says that the balls *"will go like this"* (showing the direct path going from the funnel to slot MR), after having chosen M of box II *"because it is in the middle."*

Are we to say then that this prediction of a maximum in the middle constitutes a correct intuition of probability, contradicting in this way what we supposed earlier about the lack of the notion of chance during this first stage? The prediction of the shortest path has nothing to do either with the notion of chance or with combinatoric probability (in the sense of the greatest frequency of certain combinations rather than certain other ones), but is a question of simple causality supported by the most ordinary experience. Where the question of chance and probability begins to show is when one must come to an understanding of where the balls will go when they cannot take the shortest path *"because they hit each other,"* as Eis again said, and because *"there are some on both sides."* Will they be more likely to go left or right, or are the chances the same for both sides? Such is the real problem of combinatoric probability and fortuitous distribution, and we see that it is reduced to the problem of symmetry.

An examination of the answers given by our subjects to this second question reveals a very significant fact. If the role of the direct path, in opposition to the oblique trajectories which go from the funnel to the outside slots, is usually apprehended, it is all the more striking that the factor of symmetry is absent from the predictions that children of this age make. Not only does box I not elicit a prediction of equality between the two slots, but even for box II the child does not foresee the equality for the right and left outside slots (except when he predicts an equal number of balls for all the slots). The situation with box III is even more curious: While the child predicts that the balls will go mostly to the center slot, he does not expect an equality in the two middle ones, nor an equality in the two outside ones. Predictions with box IV, finally, do not show any expectation of symmetry. It is this general lack of symmetry which tells us most about the difficulty the subjects have at this stage in imagining the shape of the whole distribution. And we can understand immediately the reason for this: To understand the system of distribution of the balls, it is necessary to realize that each ball as it falls from the funnel in a straight line in the direction of the middle slot has the same number of chances, if it encounters an obstacle, of going to one side or the other. Now, in reading the responses of Eis and especially of Fer when they are giving the trajectory of a ball which is shunted to one side (or when it hits the wall of the middle slot), we notice that the subjects do not even ask themselves about different possible combinations. They choose one or the other side without understanding that there is the same possibility of going right or left. Briefly, as we saw in Chapter I with the mixture, it is because they lack an idea of combinatoric structure that chance is not yet understood at this age; and consequently the child does not suspect the probability of a symmetrical distribution.

But there is more to it also. The third question that these observations bring up is about the generalization from one experiment to the next. Whatever the predictions of the subject, he is given, in fact, the means of controlling the prediction by the experiment, which in general furnishes him with a dispersion at once symmetrical and centered in the middle area. The child has, therefore, the means either of correcting his expectations by understanding the reasons for his error, or at least of taking into account the facts he sees to orient his later predictions. One interesting fact is that at this age there is not yet a spontaneous generalization based on the symmetrical dispersion; and this proves well enough that even when this symmetry is noted, it is not yet understood. This does not, however, mean that there is no kind of generalization yet, whether transfer or transposition. But when we see transfer or transposition they do not concern symmetry as such, and it is this fact which we need to stress.

With Man, the most primitive of the cited cases, we can discern no generalization at all yet: After seeing the balls distributed in the three slots of box II, he still believes that they will fall into a single slot of box III. With Eis, the experiment with box I has no effect on his predictions for box II; on the other hand, from box II to box III, there is some cognizance of the role of the medians and the presence of balls in all the slots, but no transfer concerning the symmetry. With Fer the equality noted for box I was probably useful in his prediction of an equality for box II. From box II to box IV, then from box IV to box III and again from box III to box IV, we note recognitions of the role of the center slot; but again, nothing concerning symmetry (which, however, we had him remark on for box IV). With Mor, there is again no generalization. As for Gon, who succeeded in making a correct prediction for the two slots of box I, he did not deduce any symmetry for box II, and his prediction of a symmetry for the two middle slots of box III (an overlap with the next stage) led him to no prediction concerning the symmetry of the two outside slots.

To sum up, symmetry is neither predicted nor even generalized following experiments done and noted by the subjects. From the point of view of notions of chance or probability, we have here a very instructive set of results. In the experiment described in Chapter I, we saw that the subjects of stage I remained incapable of individualizing the paths of the balls in the course of the mixture, and understood this mixture then as a kind of regulated process of the whole, with symmetrical movements of the two groups of mixed elements. We have now established that, because they lacked precisely this understanding of the different combinations determining the individual paths of the balls during the mixing, the subjects did not come to an understanding of any distribution of the whole based on the symmetry of the combinations in play. These two results which could seem at first glance contradictory are in reality related because they both rest on the same characteristic mental gap of this stage: the absence of schema combining the possible trajectories, with the crossing and interactions due to collisions of the elements. The drawings of the individual trajectories during the mixing, as well as the distribution of the whole are, therefore, both necessary; and it is the failure to understand their interdependence which shows that the subjects of stage I do not yet have any probabilistic intuitions.

But the facts are instructive also from the point of view of theoretical interpretations of generalization and of transfer. They tend to demonstrate that only a notion that is understood is susceptible to true generalization. In this regard, the influence of what the child noted on his later predictions concerning the role of the shortest path from the funnel to the middle slots constitutes without a doubt a true generalization. As for the tendency occasionally observed whereby the child expects all the slots to have some balls, after he has seen in the process of the experiment that this is true, this perhaps does not constitute generalization. It is enough to call it practical transfer (nonreflective) or a perceptive transposition (with anticipation), but all the intermediary steps as well as generalization are, of course, possible. It is all the more remarkable that the observed symmetry does not lead to practical transfer or to perceptive transposition. Not only is the symmetry not generalized, because of the failure to construct a combinatoric system, but it does not even lead to a simple assimilating anticipation, because of a failure to have noted that the symmetry is constantly tied to the given distributions.

3. *The second stage: Beginnings of structuring a distribution of the whole and generalization from one experiment to the next.*

Extending from about seven years of age to ten years (with a few precocious cases at six years), the second stage is characterized, in so far as random mixture is concerned (Chapter 1), by the beginnings of individualized trajectories. For the distribution of the whole group of mixed elements, the result of this is a beginning of an understanding of the structures, either uniform or centered with a symmetry of the dispersed trajectories around the center. Here are a few examples:

Ros (6; 3): Box I: After having predicted L and R alternately, he expects when all the balls are loosed at once a majority to go to the right *"because there are already two there."* (Experiment.) *"Both sides are the same."* "Do you know why?" *"No."*

Box II: *"There, there, and there."* "Most of them where?" *"Same number everywhere. No, most of them there* (M), *because that's the middle."* "And the sides?" *"Both the same. No, more on the left."* (Experiment.) "Why is it the same on both sides?" *"I don't know."* There is at the beginning then a transposition of the symmetry with a reversal next.

Box III: *"Most there and there* (ML and MR) *because that's the middle. There* (L) *fewer and there* (R) *fewer also."* "More here (R) or here (L)?" *"Both the same."* "And here (ML) and here (MR)?" *"Same too."* Thus we have generalization of the symmetry.

Box IV: correct for the most part with a central maximum and symmetry for the whole.

PER (7; 5): Box I. First ball: *"Here or there. Here* (R)." (Experiment: R.) "And the next one?" *"Here* (L), *no, there* (R)." (Experiment: L.) "And the next one?" *"There* (L), *and the next one there* (R). *I've got it."* "And all of them?" *"Here and there. The same all over. Well, I don't know. More on one side than on the other maybe."* (Experiment.) "What do you see?" *"Just about the same, because they went just about all over."*

Box II: "Where will the first ball go?" *"Inside there* (M) *because that's the middle."* "And all of them?" *"They will go just about all over, but more in the middle."* "And on the sides, more here (R) or there (L)?" *"Just about the same."* (Experiment.) *"Oh! fewer on the left."* "Why more in the middle?" *"Because the hole is there."* "And if I start over again, will there be more on the left than on the right?" *"No, about the same."* The prediction is thus contrary (with good reason) to the preceding experiment and in conformity with what was seen with box I.

Box III: A single ball will go into ML or MR rather than in L or R. "And all of them?" *"All over. Here and there the most* (ML and MR), *and fewer there* (L and R)." (Experiment.) *"Yes, that's it."*

Box IV: correct prediction for the most part, but still irregular. We start by tilting the box slightly to the right: *"Ah! maybe we moved."*

MAR (8; 0): Box I: first ball: *"It's the same. There and there. No, more likely to the left."* (Experiment: L.) "And the next one?" "*There* (R), *since there's already one there* (L)." (Experiment: R.) "And the third one?" "*There* (L)." (Experiment: L.) "And all the balls?" "*There will be the same number on both sides since there are some that roll this way and others that way.*" Thus a generalization from the results of the single trials.

Box II: "*More on both sides, and fewer in the middle.*" "Why?" "*Because there aren't enough.*" (In fact, this is a false generalization from the preceding result.) "But if we let them all come down?" "*Then the same number in the three slots.*" (We drop twenty.) "*Oh no; there are more in the middle because they go there faster. It's closer.*" "And if I let them all go?" "*More here* (M) *and fewer there* (L and R)." "More on the left or on the right?" "*Maybe fewer on the left, but it could come out equal because they roll at the same speed all over.*"

Box III: "*In all four. More here and here* (ML and MR)." "More one side?" "*Same, because they all go the same speed.*" "Why?" "*Because they're right together, those two houses* (slots ML and MR)." "And there (L and R)?" "*Fewer because that's not so close, but the same number in both.*"

Box IV: correct for the most part. Next we tilt the box without Mar seeing it: He finds the result natural "*because more came over here.*" "And if we start again, will it be like this or as it was before?" "*Like before.*"

GIS (9; 1): Box I, a single ball: "*We don't know.*" "And all of them?" "*More on one side, but we don't know which side.*" "Show me by how much." (He shows a minimal difference between the two.)

Box II, single ball: "*Here* (M) *is more certain because it's just opposite* (the funnel mouth)." "And all at once?" "*All three, maybe more here* (M)." "And the two others? . . . Can we know?" "*No.*"

Box III: ML equals MR and L equals R with the greatest number in the middle slots.

Box IV: "*There'll be more in the middle and fewer in the sides. But with the nails it's different. They'll spread them all over.*"

CRAN (9; 6): Box I: He predicts successive compensations for the single balls and a symmetry for the whole. The experiment gives a slight difference: "And if we did it all again ten times with five hundred balls, would the difference get larger or smaller?" *"It would get larger because we would always be dropping more; maybe there would be more falling to the right."*

Box II: more in the middle and the same number in L and R.

Box III: *"Anything is possible. We can't know."* But *"maybe more here* (ML) *and here* (MR)." "And the others (R and L)?" *"About the same."*

Box IV: *"Maybe more in the middle because it is steeper* (!)." But the nails *"will make them roll differently."* Without the nails: in the main a correct distribution, but still irregular. . .

LUER (10; 3): Box I: *"Maybe a few more on one side than on the other."* "What's the reason for that?" *"Because there're some balls that touch; there'll be one which will be knocked to the other side."* Thus there is a combinatoric symmetry with small fortuitous differences.

Box II: *"One or two on each side and the rest in the middle."* (Experiment.) *"Yes, I saw some hit each other as they rolled. It made them go to the side."*

Box III: an exact curve. "And if I do it again?" *"The same thing."*

Box IV: *"In all the slots because the nails will send them all over."* "And if there were no nails?" *"Fewer in the slots on the sides than in the middle."* (Experiment with the box tilted.) "Is that right?" *"No, because the board was tilted and the balls went to this side."* "You know what chance is?" *"When you play the game* boule *you think that the ball won't roll where you want, it does anyway; that's chance."*

Two essential differences mark these subjects off from those of stage I and show the appearance of the first intuitions of the notion of chance: On the one hand, there is the prediction of a certain symmetrical dispersion of the whole which is already seen as a result of different combinations or as a mixture of interacting trajectories; and, on the other hand, there is a very obvious generalization of certain observed facts when they bear precisely on this symmetrical dispersion.

Let us first of all examine this second point. We see for example that Ros after noting the symmetry of the dispersion in boxes I and II (without having predicted it before the experiment), predicts a complete symmetry in the box with four slots (III), and even an approximate symmetry in the one with eighteen slots (IV). Per generalizes for box III the symmetry noted with box I; and when the first experiment with box II proves him wrong, he even expects a correcting of the symmetry, then anticipates it for box III and the same, in part, for box IV. Mar generalizes immediately an observed fact made with box II, then applies it to boxes III and IV. Gis, Cran, and Luer give similar generalizations.

We are unable to explain this contrast with the reactions observed in stage I except by calling for an understanding of the mechanism at play: Even though the younger notice the symmetry, they do not generalize it because they do not grasp the reason for it. But the subjects of stage II do generalize it because they understand it. The beginning of this comprehension appears more and more explicitly in the words of the subjects. It is based on the gradual growth of the intuition of random mixture, which we saw in the course of this same stage in Chapter I. If the balls are dispersed in a symmetrical fashion, it is because the mixture and the collisions during their fall disperse them, sending them with the same probability to one side or the other. To explain the symmetry, the subject Per is satisfied with saying that the balls will *"go pretty much all over, but with more in the middle."* Mar is more precise: *"There will be the same number on both sides since some roll here, and others there."* Then, after he discovered that the balls piled up in the middle, *"That's because they roll there quicker, it's closer."* He justifies the symmetry in the lateral slots by adding that *"they roll everywhere pretty fast."* And, *"It's a little further, but the same in both directions."* Cran suggests the same reasoning by saying that, in the middle, *"It's steeper;"* implying that the equal slopes to the right and left give the symmetrical dispersion. Finally, Luer gives the key to what was implicit in the preceding subjects: While falling *"there are some balls that collide and after this collision there will be one that goes there and the other there"*; then, *"I saw some which touched as they were rolling. That makes them go to the side."* In short, there is a symmetry because, given the same slant for the right as for the left, the balls which mix up and collide are dispersed with as much chance to go to one side as to the other.

Because they little by little come to an understanding of the symmetrical dispersion of the mixture, these subjects often succeed in predicting symmetry of distribution by themselves, even before noting it experimentally (Per, Mar, etc.). An interesting indication is given in this regard by the reactions of the subject when, unknown to him, we tip the box to the left or the right. When the children of stage I see all the balls rolling to one side, they think it quite natural, since nothing for them is due to chance (for example, Gon in Section 2). The subjects of stage II, on the other hand, explain it immediately either by an exterior cause as distinct from fortuitous dispersion (see Per and Luer), or (like Mar) consider it to be a fluctuation attributable to chance, but say then that it will not happen the next time.

In brief, the chief characteristic of this stage is certainly the beginning of an understanding of the dispersion as a whole due to random mixture and ending up with a more or less precise symmetry with the highest frequency in the middle because of the slant. But, in contrast to the subjects of stage III, those of stage II come only to a beginning of understanding by successive approximations and without any deductive generalization beyond what was noted in their particular apparatus. In this regard it is striking to find a failure to understand the role of large numbers which we had already suspected at the same stage in Chapter I. Cran for example thinks that the small asymmetry noted during an experiment will increase instead of disappear if we drop five hundred balls ten times. On the other hand a curious reaction comes up frequently and tells the same story: The child imagines that the nails used with box IV, instead of simply reinforcing the dispersion noted in the boxes without nails, will have the tendency *"to make the balls skew,"* and send them *"all over"* to give the same distribution between the sides and the center (see Gis, Cran, and Luer). In this case as well as in the one with the large numbers, the subjects have difficulty in deducing the fact that the result of an indefinite number of combinations will result in compensation and more likelihood of random mixture. The subject thinks that each nail will skew the balls more in the direction of the sides, and does not see that this works just as well for the other direction. Thus he has not understood that the nails simply increase the mixture and the number of chance collisions, but with a general compensating effect which reinforces and does not destroy the distribution around the center. In the field of the law of large numbers also, the child who understands quite well for a small number of trials the role of compensations and their effects on symmetry, experiences a systematic difficulty in understanding the phenomenon in a generalized form because a large number of trials runs the risk, in his eyes, of reinforcing the number of skewed balls and not the compensations. In both the cases with the nails and the law of large numbers in general, the limits of this stage are marked by the inability of the subject to generalize the operations of combinations and permutations beyond the concrete situation of the experiments (which we already witnessed in the preceding chapter).

4. *The third stage: Symmetrical dispersion of the whole with immediate quantification.*

It is in general at about eleven or twelve years of age that we find responses which are immediately correct; but naturally we find some precocious subjects before this median age. Here are a few examples, beginning with a case of reactions intermediate between stages II and III:

RAG (9; 7). Box I: *"Each will come to rest in a side slot."* (Experiment.) *"One extra fell on the right."* "If I do it over again ten times, will we have a larger or a smaller difference?" *"A smaller difference."* "If I drop a hundred balls ten times, how will they fall?" *"Four hundred and ninety-eight on one side and five hundred and two in the other."* "Why will there be such a small difference when we drop so many?" *"Because only a very few will fall to one side rather than to the other."* "And if I drop only five hundred balls?" *"Because sometimes they can . . . because there is less weight* (thus less action of the balls on each other)."

Box II: He predicts that for just a few balls, they will go mostly in the middle, but that if there are many *"they will go to all sides"* with uniform distribution. After the experiment: *"They all went to this side* (R) *and to that one* (L) *at the same time, and that made many go into the middle* (gesture showing the collisions that threw them toward the middle)."

Box III: *"If you drop many balls, it will make them go mostly in those two slots* (middle) *rather than in these* (sides)." "Why?" *"When there are many they hit each other, and they go more often to the middle. Then that makes them hit each other, following each other as they fall a few more to the side and they end up there* (L) *and there* (R)."

Box IV: First he predicts the exact curve, then, *"I made a mistake, because there are nails which will make them change direction."* But after the experiment, he recognizes that the nails *"will make them fall in zigzags."* Thus we get reactions of stage III for boxes I and III and still stage II reactions for boxes II and IV.

BERG (10; 9): Box I: *"A few there and a few there. And even so maybe more on one side but not necessarily."*

Boxes II and III: correct answers.

Box IV: a bell curve *"because there are nails which will make them skew."* "Where?" *"It will probably bring them back toward the center, but according to how the nails are placed, it will make the balls go here or there* (i.e., toward the center or toward the sides)."

ANT (11; 2): Box I: *"It will be about the same on both sides."* "And if I drop one thousand?" *"There will be twenty more on one side."* "And two hundred?" *"Also twenty more."* "And fifty?" *"Just about equal—eight more."* "O.K., is it more or less regular when there are many?" *"It's less regular when there are fewer. They fall more quickly, more easily to one side."*

Box II: *"More in the center and the same on both sides."*

Box III: *"More here* (ML and MR)." "And on the sides?" *"Not many."* "More on one side?" *"No, just about the same."* "Why?" *"To have more, we'd have to have one of the slots larger."*

Box IV: *"In all the slots, but fewer on the sides."* "Any slots in which there'll be more?" *"There'll be fewer here* (center) *because of the nails. . . . Yes, there'll be some there anyway because they can hit a nail and then go in there* (gesture showing ricochet)." We tilt the box, and he notices it immediately.

KAM (11;2): Boxes I, II, and III: correct answers.

Box IV: *"They are going to hit the nails and scatter."* "Everywhere the same?" *"No* (and he sketches with his finger a beautiful bell curve without hesitating)." "Why?" *"Because the nails keep them together* (gesture of bringing them to the center)." "And without the nails?" *"They would all go toward the center."*

TAL (12; 3): Box I: *"The same on both sides, maybe not exactly, but almost."* "Why?" *"I don't know. It's chance."*

Box II: *"Just about all over, more in the middle and about the same on the two sides."*

Box III: *"More here and here* (ML and MR) *and about the same on both sides."* "Why?" *"Because it tends to go straight. They're heavy and the box is slanted."* "Is it slanted the same toward here (ML) as there (L)?" *"It's slanted less toward there* (L) *because that's a longer path."* "And what if we had a slot even further left?" *"Even less then."* "Why is there the same number of balls here and there (L and R)?" *"Because it's the same distance and the same slant."* "And why more in the center?" *"Because the ball is heavy and that pulls it in a straight line."*

Box IV: *"About the same in all the slots because the nails send them to the sides. When they hit a nail, they change direction. No, they roll like that, in a zigzag. They will fall like this* (forms a bell curve." "And if we took out the nails?" *"Just about the same way, but even so, a little different* (shows a bell curve less dispersed)." "And if we got a curve like this (right asymmetry)?" *"That would mean that they had been dropped more from the right. That would be chance."* "What if it were repeated several times?" *"I don't know—the box would be tilted, or there would be more nails on one side."*

These are the principal reactions of the third stage. After the subjects of the first stage showed that they were unable to make predictions about the distribution of the whole, because they were unable to see the compensations between the collisions and the deviations, and after the subjects of stage II showed the beginning of such an understanding, but only for small groups of elements, we see the children of stage III predicting exactly. They also justify the distribution of the balls in boxes I, II, and III, and they draw from these facts by means of generalizations a correct prediction about the most probable dispersion of the balls for box IV. This mental development goes right along with an increasingly precise ability to imagine the crossing of paths due to collisions between the balls or between the balls and the walls or nails. They also give a more thoughtful explanation about the role of the slope and about the path of the drop due to the weight of the balls.

And as usual, this stage is particularly interesting for showing developments of deductive processes leading toward the law of large numbers. With Rag already, in spite of his youth, and especially with Ant, we see clear statements of the idea that fortuitous differences diminish as the numbers increase. But the statement is connected to two predictions which are worth examining. The first is the understanding of proportionality: A difference of eight in fifty is larger than twenty in two hundred; and twenty out of two hundred is larger than twenty out of one thousand. Thus Ant is thinking of relative differences and not of absolute ones, as did those subjects who denied the law of large numbers (stage II). But most important of all, the gradual compensation expressed by the law of large numbers is explained by means of a more or less precise combinatoric scheme. For example, when there are fewer balls, Ant says, *"They fall faster, more easily to one side."* This means that without a sufficiently large mixture, there are fewer collisions and compensating ricochets. From this point of view, it is interesting to note in stage III, contrary to stage II, the child interprets correctly the role of the nails of box IV, seeing that this means simply an increase in the number of combinations with effects in both directions. Now it is precisely this operative scheme of combinations and permutations, seen in the role played by the nails, which explains the possible generalization of the effects of compensation in the case of large numbers.

5. *Uniform distribution of drops of rain on square tiles.*

After analyzing an example of centered distributions in the course of Sections 2 through 4, we shall now study a case of uniform distribution. Nothing is more common in this regard than the form of distribution which drops of rain give when falling fortuitously at the beginning of a small shower. Taking a surface composed of squared tiles, all of equal size, and the first large drops of rain which are still single and able to be counted, only certain of the tiles will be hit while others will remain dry. This constitutes a familiar physical image of chance. But as the rain continues, there is less chance that any single tile, dry for the moment, remains dry very long. However, there is an increasing probability of a regular distribution of the drops. This distribution is then uniform, contrary to the bell-shaped distribution of a sprinkler (or the balls rolling from a funnel), and it is a frequently observed fact for children at any age. If it is true, as we have just seen, that children under eleven to twelve years of age have systematic difficulty in predicting the effects of large numbers in the case of bell-curve distributions, and also in grasping that dispersions necessarily increase with compensating regularity during further trials, will the subjects not have, in this phenomenon familiar to all of them, a special chance of understanding intuitively the law of large numbers? This is what we are trying to analyze.

The technique of the experiment is as follows: A large sheet of white paper divided into squares about one inch in size simulates a tiled area, and small glass beads in the shape of cubes are dropped through a screen which is shaken, simulating raindrops or hailstones. We then have the child predict where the successive drops will fall and if all the squares will, little by little, be hit, and how the distribution will be affected by increasing the number of beads.

The results obtained are quite clear. The child at any age knows quite well that when rain falls, there are drops all over, but *all over* does not imply for small chidren a fortuitous distribution more and more regular. At the beginning, the dispersion can be regular without being fortuitous, or arbitrary without being regular; and it is only the children of stages II and III who make a synthesis of uniformity and chance, due to a gradual understanding of the law of large numbers. The intuitive recognition of the effects of large numbers does not, therefore, bring with it, from the beginning, an understanding of the process. This understanding, even for a case as simple as this one, needs the addition of combinatoric schemes.

Here are some examples of stage I:

CHAU (6; 2): *"Pretty much all over. They will fall in all the squares."* "What if there are only four raindrops?" (The child places four beads very close to each other.) "Will they fall like that? Four beads together and nothing elsewhere? Watch." (Experiment with four beads.) *"They're in more of the squares. They aren't together."* "If we continue, will it get more or less regular? . . . Will there be some squares with nothing?" *"Oh, no. In the world* (in reality), *they are all over."* "How will the drops fall?" (The child arranges the beads in a regular way, as if one had fallen in each square.) "What if it rains a lot?" *"It will get wetter."* "Will all the squares be the same if a lot fall?" *"Not just the same. Maybe there* (he points to one square) *there won't be any."* "And if we continue?" *"Everywhere."*

JOC (6; 11): *"There will be some on each square."* "Why?" *"If there are a lot, they go all over. If there are just a few, they only hit some."* "Fine, here's the rain. Three drops are going to fall." (He places them two on one square and the other on the neighboring one.) "And now fifteen drops which fall any old way." (He lines up one drop per square.) "Do you think that it will fall like that?" *"Yes."* "How many rain drops would we need to be sure that there would be some in each square?" (He counts the squares.) *"Twenty drops."* "You are sure that if twenty drops fell, they would hit the twenty squares?" *"Yes."* (Experiment with twenty beads.) "Is it as you thought?" *"No."*

DAN (7; 4) immediately does the experiment himself with ten beads, then he next predicts that the beads will fall *"there, where there aren't any because then it will all be wet."* We arrange some beads with a large variation of the number in each square: "Could they fall like that?" *"Maybe, but that's not quite right. It ought to be the same all over."* Then we take two sheets with ten squares on each one, the first with eight beads well spread out, and the second with many more beads but some concentrated on a few squares and the others spread out. "Which one is more correct?" *"The one with the most."* "The one with the most difference between squares? . . . Do you know what 'difference' means?" *"That means there are more."* "Look at these two squares (on the one, two beads, and on the other, just one) and these two (one with ten beads and the other with eleven). Which pair is the most different?" *"This one* (ten/eleven) *because this one* (one/two) *is only different by one* (one bead difference)."

We see that at this first stage the child is instinctively convinced of a regular dispersion of the rain (see for example Chau). But there are three indications that show us rather well that this understanding remains true only for the whole and does not include any analysis of the details. When it is a question of reproducing the falling of isolated drops, the child distributes them systematically as if the cloud were going to put the first drop on square 1, the second on square 2, and so forth (see Chau and especially Joc, who believes that twenty drops falling by chance are enough to hit each square for certain). And the child may, on the contrary, pile all the drops on one or two adjacent squares as if the cloud were aiming at a particular point without any total dispersion. Finally, it is clear that the child at this stage has no notion of proportionality, except for small numbers: Dan for example, sees a larger difference between the ratio ten to eleven than between one to two, and so on.

Here now are some answers at stage II which are characterized by an understanding of a progressive regularity, except, however, for large numbers:

HAN (7; 8) begins, as happened in stage I, by piling the first four drops on the same square. "And if it rains harder?" *"It will fall everywhere."* "What if we had only a few squares, where would the rain fall?" (He disperses them.) *"They would be there and there."* "And if it rains more, does it become more regular?" *"A little more. It falls in all the squares."* "What do you think of as more regular, a square with one drop and another with two, or a square with ten drops and another with eleven?" *"Ten and eleven are more regular. One and two is not the same thing. For ten and eleven, there is only a difference of one and that's not great. One and two, that's not the same thing."*

GIS (8; 10) starts out by spreading one drop per square and then disperses them. "Is it possible that there are one hundred drops here and three there?" *"No."* "One hundred and twenty there?" *"Yes."* "Is there another way that is more likely?" *"Yes, a one hundred and one hundred is more correct. It couldn't happen very often that there would be one hundred in one place and only twenty in another.*" "One and two, or ten and eleven, which is more regular?" "*It is the same thing because it's only one more in each one.*"

KAL (9; 6) says that when it starts to rain the drops cannot be concentrated in a single place. "*No, they will go just about everywhere.*" "Could we have forty here, and twenty there, and none there?" "*Yes, that could happen.*" "And one hundred on this square and none there?" "*No, that couldn't be.*" "One here and two there?" "*Yes, that's all right.*" "And fifty there and one hundred here?" "*No, because the rain falls everywhere at the same time.*" "Ninety here and one hundred there?" "*No, that couldn't be.*" "If one hundred fall on this square, how many would be on the next square?" "*There will be one hundred here and ninety-nine there.*" "Why?" "*Because in the beginning there was one and two* (he keeps thus the difference of one)."

GRA (10; 2) is a bit superior and overlaps with stage III. "If six drops fall first, what will it look like?" (He spreads three in three directions and the other three in the fourth direction.) "Will it be regular or irregular?" "*Irregular.*" "What if many fall?" "*There will be some all over.*" "After the first drops, where will the next ones fall?" "*There where there aren't any yet, there where it's dry. Since finally it will be wet all over, some certainly have to fall there where it's dry. It will be more regular like that.*" "Will it be more regular if only a few drops fall, or if many fall?" "Which way is more regular—when one falls in a square, and two fall in the next square, or when one hundred fall here, and one hundred and one in the next square?" "*Where there are more, it is always more regular.*" "And one and two or ten and eleven?" "*There is only a difference of one in each, which is the same difference, but when there are more it seems more regular.*"

The reactions of stage II show clear progress over stage I in that the uniformity of the dispersion is understood as increasing and at the same time due to chance. In fact, we find that the subjects no longer spread the drops with an artificial regularity: Gis, for example, who starts to put one drop on each square (as happened often in stage I), immediately spreads them out in imitation of chance. Thus the regularity of uniform distribution for these subjects is the result of progressive compensations, and there is a synthesis between the regular and the fortuitous, not opposition as we saw in the small children. Very significant in this regard is the refusal of the child to accept differences between squares: *"It couldn't happen often that there would be one hundred drops in one place and only twenty in the next one,"* Gis said; and Kal finds it impossible that one hundred drops fall on one square and none on the next one, as well as fifty and one hundred, or even ninety and one hundred. But he does accept differences of forty, twenty, and zero, then one of one and two and one hundred and ninety-nine. Could it be thus that the child at this stage has discovered the law of large numbers, since the compensations always seem more probable to him as the number of elements increases? In one sense the answer is yes, but only in a purely intuitive and empirical sense. What he still needs in order to grasp this process in an operative way—that is, to assimilate it as a combinatoric scheme, which is the only way that he will see the reasons for it—is the notion of the proportionality of differences. In this regard the reactions we have just seen are very revealing and give us a precise parallel with what we noted earlier about proportions in the realm of space and geometry.[1] On the one hand, the child from seven to twelve years of age does not come to a measured notion of proportion as an equality of two ratios, because he considers the differences as absolute rather than relative: Thus Gis contends that one to two or ten to eleven are *"the same thing because there is just a difference of one in each";* and Kal says that two neighboring squares will have ninety-nine and one hundred *"because in the beginning there was one on the one square and two on the other."* But the same subjects, on the other hand, already have what could be called a qualitative intuition of proportionality (based on similarities inherent in the systems of correspondences or logical multiplications). Thus Han says quite explicitly that *"ten to eleven is more regular* (than one to two) *because ten to eleven is only one drop more, which is not very great, but one to two is just not the same*

[1] Jean Piaget and Bärbel Inhelder, *The Child's Conception of Space* (New York: Norton, 1967), chap: XII.

thing." And Gra adds (also concerning one to two and ten to eleven) that "*there is only one more in each, and that's the same difference, but when there are more it's more regular.*" In brief, the increasing compensation inherent in large numbers is suspected simply because of the intuited regularity, but the decreasing differences are not yet expressable in terms of quantitative proportionality.

With stage III, on the contrary, the progressive uniform distribution is finally attributed to the decreasing of proportional differences, and not to absolute differences:

PIE (12; 5) at times still falls into reactions of younger subjects. In the presence of two pairs of squares, the one pair with one and two drops respectively and the other with ten and eleven, he declares, "*It's more regular here* (ten to eleven) *because if one and two drops fell there, then we would have twenty here since there are already ten* (thus ten to twenty equals one to two." "And if it continues to rain, will it be more or less regular?" "*It will be the same thing. There will always be more here* (ten to eleven) *and there will be one hundred and one hundred and ten because the numbers are multiplied by ten.*" "Wait a minute, it's rain we are talking about, not arithmetic." "*Then it will be one hundred and one hundred and one. That's more regular. Oh no, that's always a difference of one.*" "And if it's one thousand and one thousand and one?" "*Oh, no, it's not the same thing; it's less of a difference. When there are more, the difference counts less.*"

BIS (12; 2): "*They can fall anyplace, on one square or on another.*" "What if one hundred fall at once?" "*They will fall in a regular way because rain falls from a certain height.*" "Could, for example, one hundred fall here and forty-five there?" "*No, about the same number on each square.*" "Will the differences between the squares increase or decrease as more drops fall?" "*They will always decrease.*" "If fifty fall here, fifty-one there while one hundred fall here and one hundred and one there, which of these is more regular?" "*Here* (one hundred to one hundred and one) *because one more with one hundred is less than one with fifty.*"

These subjects understand in a new way the mechanics of progressive compensation which assures the regularity of the distribution as a function of the larger and larger number of drops. The difference between the squares *"always gets less"* or, as Bis adds, *"one more with one hundred is less difference than one with fifty."* It is this proportionalistic concept of the differences which finally gives the child's mind a hold on the combinatoric compensation inherent in large numbers.

To sum it up, the results of a particularly simple case of uniform distribution are shown to give more of a parallel than we would have thought with the stages of centered distributions: In both cases stage I showed an artificial regularity or an arbitrary dispersion (but not yet understood as fortuitous). In stage II there was an intuition of regularities (symmetrical or uniform), but without an understanding of the role of large numbers, because of a lack of combinatoric schemes in the first case and a lack of proportional differences in the second. In stage III, finally, the law of large numbers was understood in its double aspects of combinatoric and proportionalistic roles.

III. The Discovery of a Constant Relationship in Conflict with a Fortuitous Uniform Distribution[1]

To analyze the way in which the notion of chance is built into the setting of daily physical experiences, it is not sufficient to do as we have just done and study the idea of random mixture along with an interpretation of centered and uniform distributions. We still need to find out how the mind of the child succeeds in dissociating what is due to chance from what is due to the non-fortuitous in a situation in which the objects are either distributed in a uniform manner or are polarized as a function of an unknown cause. In reality it is precisely this combination of what today is commonly called uncertain and what is not uncertain (or is only weakly so) which most frequently gives rise to the intervention of the intuition of probabilities, both in the laboratory and in life.

We have already seen, moreover, in considering the bell-shaped curve (Chapter II, sections 2 to 4), an example of such a combination of the certain and the uncertain, since in the case of the balls rolling out of the funnel, the straight-line path gave a

[1] In collaboration with Marianne Denis-Prinzhorn.

simply causal relationship, and the interacting trajectories brought in the idea of fortuitous mixture. But the combination in this case is easy since the understanding of the causal relationship around which the uncertain distribution is centered precedes the understanding of the distribution. The problem which we must study now is from the other direction, starting with a fortuitous distribution and imposing in the course of the trial a causal relation which is unknown to the subject and whose existence the subject must discover. Such an experiment will presuppose induction and probabilistic judgments which will give us information on the understanding of the notion of chance at the different stages of development.

We start, thus, with a uniform distribution, that is, from a noncentered distribution, which is a function of a causal relationship. We present our subjects with a pointer which can be spun on a central pivot: On which of the equal sections of the circle which it describes in spinning will it finally come to rest? If no external force intervenes (breezes, magnetization, etc.), the points at which the needle stops will be evenly spread as a function of the sections of the circle. Since the phenomenon is well known to children (weathervanes, tops, spinning a knife on a table to see at whom it will point), we can analyze in this way the development of a new kind of uniform distribution (the example of drops of rain, Chapter II, section 5, was too elementary to exhaust the possibilities of the problem). But once the subject notes the fortuitous distribution of stopping places, we will introduce an unexpected constant element which will throw off the probabilistic predictions of the child, assuming that he actually sees the dispersion of stopping points as uncertain. We can bring in a magnet which will stop the spinner suddenly at a given section. What then will be the reaction of the subject? Will he admit that a fortuitous stopping on the same section can be repeated indefinitely, which would certainly show that he had not understood the nature of chance; or will he suppose a cause which limits this stopping by substituting for a uniform distribution a centered distribution?

1. *Technique of the experiment and general results.*

The following materials were used: The spinner is shown as a black bar painted on an aluminum disc. Below the disc and parallel with the painted bar is an iron rod which can be attracted by a magnet. The disc rests on a needle for pivot. This is placed on a square board covered with a piece of paper divided into differently colored sections, as one cuts a cake from the middle. These sections naturally have equal angles so that the chance of stopping on any one of them is the same. There are sixteen of these sections (a simplified one has eight), the colors of opposite sections being identical, there are in all eight distinct colors (four in the other board) on which the pointer can stop. (Of course the board must be lined up perfectly horizontal with a level, Figure 6.)

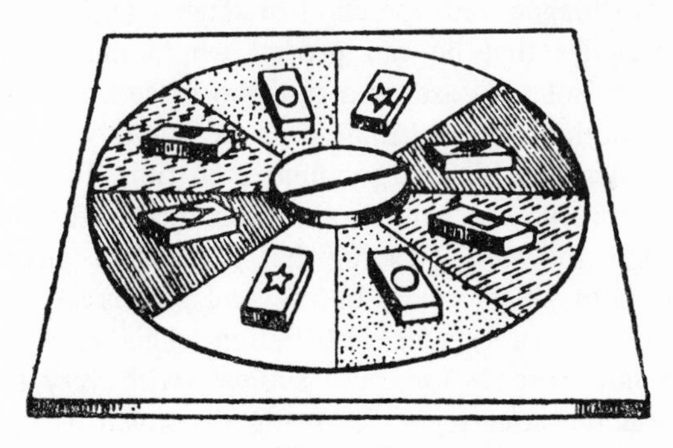

Figure 6.

In addition, during the second part of the questioning, we put eight identical boxes of matches at the edge of the paper corresponding to the eight sections of the circle. Some of them contain lead hidden in wax (weight A), others a lighter metal which is also in wax (weight B), others wax in which are some grains of metal powder (weight C), and the last ones, simply wax (weight D). Among the boxes with weight B, two contain magnets hidden in the wax. Nothing distinguishes them from the other boxes of the same weight except that the magnets will attract the iron bar and immobilize the spinner on the chosen color.

The conversation with the child bears first on the predictability and distribution of the stopping points. The child as well as the experimenter, operate the apparatus and we take careful notes on his reactions, asking, in particular, questions about large numbers: If the number of trials is increased, what will be the probable distribution of stopping points? After we have gotten his reactions on this, we next put the boxes into place, including the ones with magnets in them. When the child seeks a new explanation, he thinks generally that the weight of the boxes has stopped the black bar; we then give the child permission to rearrange the boxes (remember that those with lead are heavier than those with the magnets), to see if he will maintain his initial induction.

The reactions observed can be broken down into the three usual stages. From the point of view of uncertain dispersions when the spinner is not magnetized, the child of stage I (up to seven years of age) imagines that he can predict where the bar will stop, either in each isolated case or a function of the preceding cases, but with no notion of a distribution of the whole which would increase in regularity with a large number of trials. In the course of stage II (from seven to ten years), the subject refuses to predict where the bar will stop, guessing that it could be anywhere if it is a question of single spins, but several successive trials allow him to anticipate a regular distribution. This regularity, however, does not increase for these subjects with very large numbers, but, on the contrary, risks being destroyed as we increase the number of the trials. In the course of stage III, finally, the uniform distribution is tied to large numbers.

When the bar is magnetized, the child of stage I limits himself to noting what he sees, sometimes with justifications after the fact, but without a coherent explanation and especially without a clear difference from what he said when he was faced with simple chance. In the course of stage II there is progressive induction, with some beginnings of generalization, and an understanding of the difference between fortuitous dispersion and causal regularity. Finally, in stage III there is a search for and the discovery of the controlling cause after the trial, and a clear intuition of the probabilities.

2. *The first stage: I. Illusory prediction of isolated results.*

It is usual for subjects of stage I not to admit the equivalence of the different possible stopping places and, therefore, to attempt to guess where it will be. A large number of questions are asked concerning the formation of the notion of chance and, in general, about the point of contact between thought and reality. First, does the child really believe in the possibility of predicting where the bar will stop? Outside of playful attitudes which tend to arise when a problem goes beyond the mental level of the subject, it does seem that he adopts a nuanced attitude: He knows quite well that he is not likely to be able to predict the color on which the bar will stop, but he does believe in the legitimacy of such a prediction and tries to guess the result. On what then does he base his guess? The reactions of this first stage do have a certain interest. The child oscillates quickly between two solutions which attract him in varying degrees according to results obtained in the trials. Either the bar will have the tendency to come back to a color on which it has already stopped, or it will, on the contrary, stop on the colors not yet touched. As for the first prediction (before the first trial), or for a choice of which of the colors not yet touched, different motives are at play: the choice of a preferred color, the order of the color (for example, blue coming after red), the force with which one spins the disc, the ability of the player to control the spin to get the desired color, and so forth (note that the disc turns easily and makes a number of spins, thus making any control impossible). But we are not concerned with the reasons given—it is only the mechanism of thought that is interesting. In this regard, we see that the attitude of the child alternates between two poles whose relationship we seek to understand: a certain phenomenon which consists in tying anything to anything else (whether in advance of the experiment or because the experiment brought them together), and a certain egocentrism which consists in interpreting the relationships as functions of the activity itself.

Here are some examples of these several reactions:

JEA (3; 10): "Do we know where the bar will stop?" *"On the red."* "Is that certain or not?" *"Certain."* (We do the experiment: green.) "Where now?" *"On the yellow."* "Why?" *"Because that's a wonderful yellow."* (Experiment: blue.) *"Oh, no, on the blue."* "Where now?" *"Red."* (Experiment: red.) "And next?" *"Again red."* "Why? . . ." The same kind of answers, without motives. We put the magnets in place: *"The bar got tired."*

NOU (4; 11): "Can we know where it is going to stop?" *"No, because if we said it was going to stop on blue, and it went by the blue, we wouldn't know."* "But do you have any idea?" *"I think it will be on the blue."* (Experiment: green.) "If we do it again?" *"The blue."* "Blue or yellow?" *"Blue because it will go by the yellow."* (Experiment: green.) "Where has it not yet been?" *"On the blue and the yellow."* "Which one of those two is more likely?" *"Blue because it'll go by the yellow. It goes so slowly at the end."*

CHAP (5; 4): "Can we know where it will stop?" *"Yes, on the red."* "Is that certain?" *"Yes."* "Quite sure?" *"Yes."* (Experiment: blue.) "And the next time?" *"On the red this time for sure."* "Why?" *"Because I see it there already* (in his mind)." (Experiment: green.) "Did it land on red?" *"No, on green because that's next to red."* "And the next time?" *"On the blue because for me that's the same thing* (thus he considers blue subjectively as the same genre as green)." (Experiment: green.) "Can it stop anywhere?" *"Yes."* "If we do it again, will it be more likely green or another color?" *"Another color because it can't always go to the same color. Already it has stopped many times on the same one."*

MON (5; 3): "Can you tell me where the arrow is going to stop?" *"There* (green)." "You're sure?" *"No."* (Experiment: rose.) "What if we do it again?" *"It will stop on another color."* "Which one?" *"Blue."* "Can we be sure?" *"Yes."* We spin it, and Mon then says, *"I believe that it will stop on another color."* (The bar stops on brown.) *"Yes, I knew it."* "And now?"

"Yellow." (Experiment: yellow.) *"I knew it."* "How did you know it?" *"Don't know."* "And now?" *"On another."* "Could it happen that it would go twice to the same color?" *"No."* (Experiment: green.) "Has it already stopped there?" *"Yes."* "Could that happen often?" *"No. It will go where it hasn't been."* We continue and Mon maintains that each time it will stop on a new color, in spite of the results of the experiments, and she thinks also that the adult knows where it will stop.

FIS (5; 8) believes, on the contrary, that after the first experiments, which he does himself, the bar will stop on the same color *"as the first time."* "Why?" *"Because it goes very quickly to the same color."* He then explains every mistaken prediction by a fault in the spinning. *"It will land on green."* (Experiment: blue.) *"It's because we didn't turn it fast enough."*

BER (5; 9): In addition to the speed of the spin, he invokes the order of the colors: "Do you think that we can know ahead of time?" *"No."* (Experiment: white.) "And now?" *"On another color."* (He points to the following ones.) *"I did it a little hard."* (The bar lands on green.) "And if we do it again?" *"On the black, because that's a little further."* "What if I turn it with the same force?" *"It'll land on the same color."* "Do you think you could spin it with the same force?" *"No."* "And me?"

ED (6; 9): "Can we know where it is going to stop?" *"On the blue."* "And if I say red and you say blue, who will win?" *"Me. It'll always be blue. You can see that by the speed. You can see that it wants* (equals 'is going') *to go there."* "Are you sure?" *"Yes."* (Experiment: brown.) "Were you right?" *"It's a little hard. I'll spin it myself; it'll hit rose."* "Do you know better when you do it yourself?" *"I believe so."* (Experiment: orange.) "And the next time?" *"On black."* "Sure?" *"Yes."* (Experiment: rose.) *"Now it has to hit the white."* (Experiment: orange.) "Does it do what you say?" *"No."* (Successive experiments: brown, then white.) "This time on the same color or a different one?" *"On a color where it hasn't been. No, it will land on white again."* "Now, look, has it been on all the colors?" *"No, not on green."* "Why?" *"Because I didn't spin it right. Later it'll go; it has to go at least once to green."* "Could it happen that it never lands on green?" *"It can happen, sometimes. It doesn't want to."* (Experiment:

yellow.) "And now?" *"On green."* "You're sure?" *"No."* "Can one be sure?" *"Sometimes, when you're very sure."* "Can I be?" *"I think so."* "How?" *"You can stop it."*

WAG (7;2), the same: *"On the white one."* "Why?" *"We know from the speed that it'll get there."*

The element that is common to these diverse reactions is that predicting, at least in theory, where the bar will stop is possible. On the one hand, in spite of the large number of turns of the disc, the child imagines that it is possible to get a certain color by controlling the speeds (Fis and Ber), and even to stop the disc as it turns (Ed). But on the other hand, the child attributes a certain spontaneity to the movements of the disc, or says that it is *"tired"* (Jea), or that it *"does not want to stop"* on the chosen color (Ed). There is, thus, for the subjects at this stage the possibility of predicting in theory even if it is not possible in fact. It is so in theory, they think, because one ought to be able to count on the correct spinning of the disc, as if the twenty or thirty spins it describes before stopping did not depend on fortuitous fluctuations. But it is not predictable, in fact, because of errors of control due to the player or to resistances in the apparatus. Chance is thus replaced by the arbitrary explanation of invisible causes.

But there still remains the possibility of guessing. Thus Nou, after declaring that *"one cannot know,"* next affirms six times in a row that the bar will stop *"on blue,"* in spite of the fact that the experiment comes out differently each time. What then are the motives invoked in making predictions?

The most primitive reactions are those characteristic of substage I A (Jea and Nou), who simply guess the color with no plausible motive and who may change their prediction when the experiment does not confirm it, or who even repeat it right to the end (Nou: *"it will be on blue"*), no doubt because it is a favorite color. Chap shows clearly the mechanism of this kind of anticipation by the mental vision, *"on red this time for sure because I already see it there."* (It is in no way a perception of the observed situation.)

But starting from their initial reactions, we observe a particular kind of relationship between one result and the following one: A color will be predicted because it was the one just landed on. Thus when Jea saw the bar reach red, she predicted red for the next trial. This repetition, however, cannot be considered as a beginning of an ordering of the distribution of the whole. It is only a phenomenist indication which guides the isolated prediction.

On the contrary, with the final reactions characteristic of level I B we notice an effort to grasp the relationships of the successive results. First we must cite three forms of relationships which remain rather primitive. The first is the tendency to predict colors according to their similarities. This is what we found with Chap, who related blue with green (*"because for me it's always the same thing"*). A second form is deciding that the sequence of stops ought to correspond to the succession of colors on the sheet. For example, Ber predicts *"on black because it's just a little further."* It is clear, however, that the bar could stop in this order only if it were controlled by the manner of spinning. And the third relationship between the trials ties the stops to the speed of the spinner. Fis, Ed, and Wag express a direct causality between the stopping points and the actions of the player: *"It's a bit difficult"* (to guess the results of another's trials), said Ed. *"I'll spin it myself. It'll land on rose."*

It is obvious that these three primitive forms of relationship are still negations of chance, even though the search for an explanation of the whole sequence of trials ought sooner or later to put the subject on the trail of a statistical law. To attribute to the bar a tendency to take into account similarities between colors or their order of succession is to deny the distribution of the fortuitous stops. On the other hand, to believe that one can aim for a particular color in spinning the disc, is to neglect the interaction of multiple causes which are in play during the rotation, and also to replace the uncertain distribution by the hypothesis of a simple causality.

The last form of relationships of the results brings up, however, a notion which will become increasingly important in the course of later stages. The child realizes very quickly, in fact, that the color to predict is the one not landed on. Chap says, for example, that the bar will stop *"on another color* (one not yet touched) *because it can't always go to the same color."* And we find this same idea with Mon, Ed, and many others. Now it is clear that this notion of compensation is likely to lead to an intuition of frequencies such as found in a uniform distribution (as we began to see in Chapter II). But at the stage now being considered, this notion of compensation does not go any further than the already noted relationships (resemblance, order, or even simple repetition). That is, it remains at once more elementary in its logical structure than an intuition of uniform frequencies, and more complex because of the multiplicity of factors which enter.

First of all, the notion of compensation is probably influenced by experience, because it is, in fact, rare to see the bar stop several times on the same color. But experience is certainly not a sufficient explanation for the attitude of the subject: He has too little respect for facts to come to such a clear conclusion. No doubt there is some influence of an element of compensation that arises in the child himself. Midway between the perception of symmetry, the intellectual scheme which will lead to the principle of sufficient reason (this is illustrated by the fact that one color is no more favored than another) and the moral idea of justice, this notion of compensation is no doubt inspired by a spontaneous reciprocity which influences the first individual actions of children.

Taken as a whole then, the reactions of this stage show either isolated predictions thought of as valid (I A), or the beginning of an ability to relate the successive results, but without any notion of a distribution of the whole (I B). In both cases the motives for the predictions are at the same time phenomenistic (repetitions or changes noted which influence the next predictions) and egocentric (confidence in one's own choice, subjective resemblances, and especially a belief that the spin can be controlled), and different weight is given to each of these coexisting elements. These diverse reactions have in comomn a kind of implicit negation of chance because of the failure to understand the mixture of influences in the play of the phenomena. The subject does not yet have the attitude of a gambler, but tries rather to guess by postulating, in whatever form, a hidden connection between the successive trials, and he bases his interpretation of the isolated results on this hidden connection.

3. *The first stage: II. The reaction to the magnet.*

The best criterion for an understanding of chance is surely to be sought in the distinction between uniform dispersions and those with a constant influence which excludes any random dispersion by centering it at a precise point. It will be useful, therefore, to analyze carefully how the subjects of stage I react to the immobilization of the bar by the magnets: Will they see in this new phenomenon the contrary of what happened before, or will they make some relationship between the two, saying that this is a particular case (not seen as fortuitous) of the arbitrary distributions noted up until now? But the younger subjects of this stage (level I A) do not show the same surprise: The new set of observations are accepted as the first ones. In the course of the next stage (I B) there is, on the contrary, growing surprise, but with as many explanations as particular situations to be interpreted:

JEA (3; 10) sees the bar stop in front of the boxes and says, *"It got tired."* Next he tries to push it on and sees it turn back. But what he sees is just as natural to him as the preceding unexplained stops.

NOU (4; 11): *"It stopped at the box of matches."* "Why?" *"It stopped on yellow."* (The boxes do, in fact, have the top side yellow and Nou connects them with the colors on the sheet.) "And now where do you think that the bar will stop?" *"Always in front of the boxes."* "Do you know why?" *"Yes, if we change the place of the boxes, if we put them by red, it will stop on red."* "Take a try." (He does the experiment.) *"Yes."* "Why?" *"By chance."* Note the unexpected use of this term to explain what is precisely regular.

CHAP (5; 4): *"It stopped at the match boxes."* "And if we do it again, will it stop at the same place or elsewhere?" *"Probably elsewhere."* (Experiment.) "And now?" *"Like that, at the match boxes."* "Is there any reason, or is it chance?" *"It's chance."* "If it stops ten times in a row in front of the boxes, is it chance or is there a reason?" *"It's chance."* (We do the experiment four more times.) "What if we put the boxes next to red?" *"It will go to red."* (Experiment: He accepts the result without surprise.) "Is there a reason?" *"Those boxes, they kind of hold on to the needle."* "Why?" *"It's chance."*

Mon (5; 3): The boxes with the magnets are on yellow: *"Again on yellow."* "And the next time?" *"On green because it hasn't stopped there yet."* (Experiment: yellow.) *"Again on yellow!"* "Now?" *"On green. Green's the last one not touched. If it doesn't go to green, I give up."* (Experiment.) *"What's that? Again yellow!"* "Next time?" *"Green. Too bad, go for the green, you."* (He addresses the bar.) (Experiment: yellow.) "Now if we put the boxes next to green, will it stop on green?" *"Yes."* (Experi- "If we do it again?" *"Again on green."* (Experiment.) *"Yes I knew* it." "What if we switch the boxes? (putting empty boxes next to green)" *"Still on green."* (Experiment: yellow.) "And now?" (The empty boxes are still on green.) *"On green."*

Fis (5; 8), whom we have seen explain the stopping points by an intentional regulating of the speed of the spin, is not at all surprised that the bar stops repeatedly in front of the boxes with magnets. *"Yes, it's the same thing because it was thrown with the same sort of speed."* "But before, why didn't it stop at the same place?" *"Because you didn't spin it fast enough."* (See Section 2 for the beginning of Fis's interview.) "And now is there a reason for it to stop in front of the boxes, or is it by chance?" *"A reason."* "Because of the boxes or the bar?" *"Because of the bar."* (We take off the boxes and the bar stops elsewhere.) *"No, because of the boxes."* (We put the boxes back and the bar again stops in front of them.) *"It's because we spin it with the same speed."*

Ber (5; 9): The magnets are on black: "Did you notice anything special?" *"The bar is on black."* "And where will it stop now?" *"On red."* (Experiment: black.) "Did you see the way it stopped?" *"No."* (Experiment is repeated.) *"It shakes back and forth like this* (i.e., the bar oscillates)." "Did it do that before?" *"Yes."* (We take the boxes away.) *"No."* "Watch." (We put the boxes on yellow.) *"Now it won't stop on black."* (Experiment.) *"It's doing it* (oscillating). *Yes, on yellow."* "What am I using these boxes for?" *"They grab hold of that turning thing."* "How?" *"Don't know. I can make myself a thing like that."* (Next experiment.) "Why does it hold the bar?" *"Because it's not straight, not quite straight."* (He believes there is a deviation of the trajectory in front of the boxes. We do the experiment again.) *"I don't know why that all the time stops right where the boxes are."* "Let's try it with these boxes (without the magnet: We do the experiment and the bar does not stop in front of the boxes)." *"Are they full? I want to see what's in them* (sees the wax). *Oh, it's that; but this one is lighter."* "And so?" *"It doesn't work."* "What if I take these heavy ones (the same weight as those with the magnets, but without magnets)?" *"It'll go to the boxes."* "Watch (experiment)." *"No."* "And these (same weight but with magnets)?" *"Yes* (experiment)." "Why doesn't it do it with the others?" *"Because they are lighter."* "But this is the same weight." (He compares the weights.) *"Yes."* "Then why doesn't it work with them?" *"Because they're the same weight."*

ED (6; 9) predicts something as soon as we place the boxes. "Where will the bar go?" *"To the boxes."* (Experiment: He is obviously enchanted.) *"It's the boxes that make it stop, makes it heavy."* "And now?" (We put the boxes on another color.) *"Yes, it'll go there again because it makes it heavy, makes it stop."* (Experiment.) *"Yes."* (Light boxes: Bar does not stop in front of them.) *"Oh, I didn't spin it fast enough. No, it's because they're lighter, and don't make it stop."* (Next we place the heavier boxes, no magnets.) "Where will it go now?" *"To the heaviest ones."* (The bar does not stop.) *"Oh, no, it's the middleweight ones that stop it."* (Now we put the heavy ones, the medium weight ones with magnets, and the light ones.) "Now where will it go?" *"To the heavy ones, no, to the light ones."*

CLAU (6; 11): Magnets on white: *"It's because there's something that stops it."* "Where?" *"In the machine* (in the spinner)." "And when there are no boxes, why does it work?" (Experiment.) *"It's the boxes, they touch the machine* (he does the experiment again)." *"It's the boxes that turn it. The point* (of the bar) *makes it stop."* "How did you see that?" *"It's the boxes that make it turn."* "But it's you who made it go around." *"Yes, but the boxes stop it. It's the white wood* (along the sides of the match boxes) *which stops it."* (We open up the match boxes.) *"Oh, wax, that's heavy."* "What if we use these light boxes?" *"Those'll do the same thing. They're a little heavy."* (Experiment.) *"Oh, no, not that time. We have to put a lot of boxes around. The bar will go to the heavy ones and stop."* (Experiment with the magnets.) *"Yes, because the heavy ones make it stop, make it shake back and forth* (oscillate)." "And now (we have put back the heaviest ones in addition to the ones with magnets)?" *"It will stop by the heaviest ones."* (Experiment.) "Why not?" *"Because the heaviest ones sent it back to the less heavy ones."* "What sent it back?" *"The boxes are heavy; that makes things move; that makes it lean a bit."* (We remove the heavy ones, with and without magnets.) *"That time it didn't make it go back. I didn't turn hard enough. When I turn it hard, it makes it go there and then there."* (He does it again.) *"Oh, I don't know why. Each time I turn it, it first goes there and there* (he points to successive colors), *and when I turn it hard it goes toward the boxes."*

WAG (7; 2): *"It's because of the boxes. The bar loves to stop there. It's the speed that makes it go there."*

The comparison of these results with those from Section 2 observed with the same subjects is very instructive from two points of view. On the one hand, it allows us to judge differences of attitude in the subjects (or the absence of any differences) who are faced with a fortuitous distribution and a constant relation. On the other hand, it gives us an especially good opportunity to identify processes of reasoning used by the subjects to establish a table of variability and also to induce what is properly called a law.

From the first point of view, during stage I we observe a progression of attitudes which is the same when the subjects are faced with a constant relationship as when faced with an uncertain distribution. At level I A, the child is limited to noting without surprise the stopping of the bar in front of the boxes and to reasoning on this particular fact in the same way that he did when the bar stopped anyplace during the fortuitous distribution. Thus Jea says no more than that the bar stopped because *"it is tired,"* and not because *"the boxes make it stop at yellow,"* as if a particular color could make the bar stop. Nou foresees quite well that the phenomenon will be repeated if we put the boxes on red, but this is because the boxes attract the bar, as does a particular color (like blue, in the first part of his interview, Section 2). Chap, who is in transition to level I B, recognizes that the bar stops in front of the boxes, but thinks that the next time it could happen differently. Even if this is repeated ten times in a row, it would be due to chance, that is, due to a particular but arbitrary reason such as, *"these boxes make the bar back up a little. That's chance* (Nou used this term in the same sense about a regularity, apparently without reason)." Mon, who quite obviously is on level I B and who foresaw during the fortuitous distribution a new color for each trial, keeps the same system of prediction after noting that the boxes attract the bar to yellow: The bar will next have to go *"to green because it hasn't been there yet."* And he says this without taking into consideration the placement of the boxes. In the same way Fis relates the stopping in front of the boxes in his first explanation with the speed of spinning: The bar stops *"because of the boxes; it's because we spin it at the same speed."* Wag, in spite of being seven years old, answers the same way.

To sum up, when the bar stops constantly in front of a particular matchbox, none of the subjects calls this a new kind of fact, different from the fortuitous stops on any color. This is because the subjects have no notion of chance and yet are forced to give a reason to each of the stops which they do not consider as fortuitous. As a result they find it quite natural that the bar stops in front of the boxes. This regular stopping, whose cause they do not understand, seems to them to be of the same order as the earlier stops which were apparently irregular but to which they assigned a hidden regularity. What Nou and Chap call chance is precisely this kind of regularity, suspected yet remaining unexplained, and for which the spontaneous Why? of children seeks the reasons, and which adults do not succeed in answering clearly! [1]

It is only at about six to seven years of age, with Ber, Ed, and Clau, in the second half of substage I B, that we see an attitude of surprise appearing, as well as the feeling for a heterogeneity between the fortuitous distribution observed until now and the constant relationship which will take over from now on. But are we to conclude, therefore, that these subjects possess the notion of chance better than their responses seemed to show in Section 2 and can distinguish uncertain distributions from simple causal

[1] See Jean Piaget, *The Language and Thought of the Child,* 3d ed. (New York: Humanities Press, 1959), chap. VI.

laws? In reality, even though these children are somewhat superior to the earlier ones, they failed to understand the uniform distribution of the fortuitous stopping points during the first part of the interviews (Section 2) because they constantly sought regular causes hidden beneath chance (order of colors or the speed of the spin, etc.). Even after they see the existence of a constant relationship, they fail to establish it inductively, without contradiction. (They clearly cannot explain the situation since the magnets are not visible.) Now, looking at this more closely, we perceive that the two failures come from the same reasons.

It is here that we need the second point of view mentioned at the beginning of this discussion: the common reasoning processes used by the subjects to construct a distribution, or a table of variability, and to understand the constant relationship through induction. To grasp the notion of chance, one must be capable of induction, since an uncertain distribution, like simple causal relationships, is to be constructed by means of experience and it is only by knowing these relationships that they will discover, by contrast, fortuitous distributions. It is, therefore, not without reason that our subjects miss both the discovery of uniform distribution of the stopping points while the disc is spinning at random and also the induction of the constant relationship when only certain boxes immobilize the bar. But to combine the several successive trials during the uncertain distribution as well as to tie together the successive observations of a constant relationship which are given by the magnets demands that the subjects have logical and operative instruments indispensable for all reasoning. And this is what our present subjects lack. In other words, we find here in a new form what we observed in Chapter I: To grasp the irreversibility of chance, one must be capable of combining reversible operations. It is because they lack such operations that Ber, Ed, Clau, and the others do not get to the point of grasping the total distribution of the fortuitous variations, nor to an induction of a causal relationship. In neither case did they succeed in reasoning from the facts nor in being able to extract from them relationships which could be grouped together.

It is for this reason that Ber thought that it was a deviation of the trajectory that stopped the bar and was surprised by the difference of weight. From this he made quite a legitimate hypothesis according to which the lighter boxes did not affect the bar but the heavy ones stopped it. We then gave him boxes of the same weight as those with the magnets and asked why they had no effect on the bar. He answered that it was *"because they were of the same weight,"* without noticing the contradiction. For him this meant that the exception did not affect the law. In the same way, Ed explained the stopping of the bar by the weight of the boxes, and then added that neither the lightest nor the heaviest had an effect on the bar, concluding that only middleweight boxes were effective. Must we then attribute to him the hypothesis (already somewhat bothersome) of the necessity of an optimum weight? Not at all, for immediately afterward he made a prediction as if there were a return to a fortuitous distribution, saying that the bar would stop *"on the heaviest . . . no, on the lightest ones."* But Clau beat all records with his absence of a coherent generalization: The boxes stop the bar *"because of the white wood"* along the sides, then because of their weight, but the heaviest box *"makes the bar go back toward the lightest."* Afterward, when we take away the magnets, the remaining boxes do not work because we did not *"spin hard enough."* Briefly, for any number of situations, many factors are brought in, with no exclusions, no generalizations.

We can understand the uniformity of reactions at this stage: It is because of the failure to know how to tie together logically successive observations that the child does not come to the induction of a cause nor to the discovery of a uniform distribution of the fortuitous stops. Certainly we should not expect children of four to seven years of age, even if they did think logically, to discover magnetization as the cause of the stopping of the bar; but they could come to the point (and they will to a certain extent in the next stage) of rejecting the hypothesis of the weight of the boxes as being the effective action, since even this hypothesis has contradictions with the observed facts. And they could also come to the recognition that only two specific boxes stopped the bar. Now, the interesting fact is that, lacking the necessary logical operations, the subjects of stage I do not succeed in excluding the false hypotheses or the imaginary explantions, either in the realm of causal order or in random mixture. In other words, since chance is in opposition to order, the subjects who are not yet in possession of the operations necessary for discovering a true causal order are also not able to recognize the existence of the fortuitous. Everything seems to them on the edge of an order which is suspected but not established, of an arbitrary nature, not uncertain but simply capricious.

4. *The second stage: I. Gradual establishment of successive fortuitous results.*

The subjects of stage II begin by doubting the predictability of single trials, and then come to see more and more clearly the notion of a distribution of the whole from the successive stopping points of the bar.

DAN (6; 8): "Can we know where the bar will stop?" *"I don't know."* (Experiment: rose.) "And now?" *"We'll have to wait and see."* "On rose again?" *"No."* "Where then?" *"I don't know."* "Can we know beforehand?" *"No."* (Experiment: orange.) "What if I ask you to guess?" *"I don't know."* "Could it land on a color where it has already been?" *"No."* "Never?" *"It could happen."* "Often?" *"No."* "And now?" *"Don't know. It would be funny if it hit red again or orange."* "What's more likely, one of those two colors or another?" *"One that hasn't been hit yet."* (Experiment: green.) "And now?" *"Brown or white."* (Experiment: brown.) *"That wasn't quite right; that was just a guessing game."* "And this time?" *"White or yellow, or maybe again on brown."* (Experiment: rose.) "And now?" *"All of them."* "Do adults know?" *"No, only God."* "How many colors are there that it hasn't hit yet?" *"One, white."* "Might it happen that it never hits white?" *"Yes."*

PER (7; 10): "Can we know where it will stop?" *"No."* "Never?" *"No."* "Guess." *"Yellow."* (Experiment: red.) "Can we know?" *"No."* "Where do you think it will go now?" *"Maybe blue."* "Do you think that it'll stop always on a different color?" *"Yes."* "Does it ever happen that it would stop at the same place several times in a row?" *"Yes."* (Experiment: yellow.) "Does it happen that it hits the same color four times in a row?" *"Maybe."* Etc.

DESP (7; 6): Same reactions: "When we spin it eight times in a row, is it more likely for it to fall on the same color, or for it to hit all the colors?" *"It is easier to predict it when we do it often because it can stop at different colors many times, and it won't stop at the same color many times."*

ANK (7; 1): "Can we know?" *"No, it turns and the speed gets smaller and smaller and we can't know where it is going to stop."* "And this time?" *"On one of the other colors. That's more likely because it can't stop always on the same ones."* "If I spun it ten or twenty times, could there be one color at which it never stopped?" *"Yes, that could happen. That would happen more often if we did it only ten times rather than twenty."*

TOR (8 ;10) is an interesting case because he has used chance several times in a row in spite of his initial hesitation. "Could you guess?" *Brown."* "You sure?" *"Not completely."* "Why?" *"One is never completely sure, as for everything."* "What if I asked you how much are one and one?" *"Oh, yes, I'm sure. That's two. I'm certain."* "And can I know what color?" *"Oh, no."* (Experiment: brown.) "And now?" *"Green."* "Sure?" *"No, not sure."* (Experiment: green.) "You're amazing. And now?" *"Rose."* "You're sure?" *"No."* (Experiment: rose.) "And now?" *"Yellow."* "Sure?" *"Yes."* "Why?" *"Because before I've been always right."* (Experiment: brown.) *"I lost. Now, black."* (Experiment: rose.) *"Lost."* "What if I asked you to guess again?" *"I would say a color which hasn't been hit."* "Do you think they'll all be hit?" *"Yes, maybe."* "Where will this one go?" *"Yellow."* (Experiment: brown.) *"Lost! already five times it's hit brown."* "Do you think that if I do it tomorrow, it will hit brown five times?" *"Oh, that depends. It'll maybe be different because of the bar and how it turns."* "Imagine that I did it sixteen hundred times. That would be two hundred times on each color if it happened regularly. Do you believe that it would be more or less regular than it has been today?" *"The differences would be less. But I count like this: five hundred* (he multiplies the five brown hits by one hundred), *three hundred, two hundred, that gives more* (he multiplies the given differences by the same number, one hundred, and finds thus that they become greater)."

WEB (8; 4): "Can you guess?" *"Not sure. It depends on where it will stop.*" "Would you say more likely on new colors?" *"Yes, where it hasn't yet stopped."* "Sure?" *"No, it depends on where it will stop."* "Will an adult predict correctly?" *"No."* "Could it fall twice on each color?" *"Yes, but it could also happen that it stops several times on the same one."* "If we did it many times, would it be more regular with few colors, or with many?" *"Easier with few."*

DERB (9; 6): "Guess." *"Blue."* "Sure?" *"Not very. It could stop on another color."* "You say on colors which have been hit already?" *"No, on those which haven't been hit yet. Green."* (Experiment: green.) "How did you know that?" *"That was chance."* "If we played all afternoon, would you know better then?" *"No, we couldn't. We're never completely sure."* "If we do it sixteen times, could it fall twice on every color?" *"No, that'll happen sometimes, but not all the time."* "And if we did it one hundred and sixty times, would that mean twenty times on each color?" *"No, that's not at all possible."* "Would the differences be larger or smaller?" *"It would be less regular."*

FLUM (10; 11): Same reactions. "And now?" *"I don't know at all. I don't want to say anything ahead of time. We'll see in a minute."* "And if we continue?" *"They haven't all been hit, and it could happen."* "We've done it sixteen times. What if we did it sixteen hundred times?" *"It would be the same. We'd just add zeros."* "Do you really think so?" *"It depends. We can't know."* "Would it be more or less regular?" *"Less regular."* "Why?" *"I can't explain, but I understand what we've done: It's more regular here when we do it less."*

FAR (11; 8) also thinks that the regularity diminishes with increase of number. "What's most unusual: about one hundred hitting each color with eight hundred trials, or two on each color with sixteen trials?" *"One hundred out of eight hundred."*

CHEN (11; 11) is not sure that each color will be hit with only a few trials, because *"we would have to spin many times."* "And then would it be certain?" *"Oh, no, we'd have to spin it twenty-five times maybe."* "And then would it be quite certain?" *"Oh, no, maybe even thirty times, and even then I wouldn't be absolutely certain."* "Which is more probable: two on each color with sixteen spins, or about one hundred on each with eight hundred?" *"More likely with sixteen spins and two on each color because two is less than one hundred. When there's chance, its easier."*

We see the significance of these facts which give a complete gradation of the progressive structure of relationships of distribution, and which thus indicate the acquisition of the notion of chance.

The initial reaction of this stage is, in fact, to refuse to predict the stopping places of the disc when done singly, that is, independently of the whole. The child is limited to saying, "I don't know" or "We have to wait" or "We'll certainly find out." But this negative attitude hides two clear points of progress over stage I. The first is the comprehension of the fact that the stopping of the bar is fortuitous and that there is no hidden regularity to be looked for in the phenomenon. In other words, the child begins to substitute the notion of chance for caprice, while in stage I everything was conceived of as regulated but according to hidden and arbitrary laws. The subjects just cited know, as Tor says, that *"we are never completely sure, as for everything!"* He says this because of the interaction of factors (*"because of the bar and also how it is turned"*). This subject Tor is particularly remarkable because luck gave him results three times in a row confirming predictions which he himself did not believe. For a moment then he wondered if there were a hidden trick, but, with his first error, he regained his initial prudence. Briefly, the child no longer believes that isolated trials can be predicted, but the second aspect of his progress is that this is compensated for by his predictions of the whole which relate to the distribution of the points of stopping and no longer a prediction of each trial taken alone.

We see this first in the new significance given to the intuition (already seen frequently in stage I) that a color not yet hit has more chances of being hit in several successive trials than one already touched. But if there is still some residue of compensation in this judgment (perceived symmetry or equalization), it is obvious that the emphasis is becoming probabilistic since the child is surprised by a frequently repeated hit on one color, even though he recognizes it as logically possible (see Tor for this also). From the age of seven onward we find an intuition of frequencies or regularity of distributions and, consequently, rarity of irregular combinations: Certain repetitions appear to be more unusual to the subject than are the alternations which, themselves, are all the more frequent as the distribution appears more fortuitous.

We must note, however, that, as this intuition of frequencies is presented here, it constitutes only a new aspect of that intuition of random mixture which we thought that we saw in Chapters I and II as the beginning point for the physical notion of chance. It is, in fact, to the extent that the successive spins of the disc are independent of each other, that is, showing variety of speeds, and following each other with no order, that the repetition of a color is least probable (see explanation of Ank for reasons for the unpredictability). In other words, this is because, in their eyes, the whole group of spins is made up of a number of successive trials, independent and differentiated by variable speeds, and the children then make the judgment that a repetition on the same color is improbable and the irregular alternation is probable. This is why they no longer give as reasons for their predictions the order of the colors or the speed of the spin. We understand retrospectively, therefore, why in stage I each single spin was thought to be predictable: This was because there was no interaction or mixture, but rather an imagined hidden connection (order, speed, resemblance, etc.) between the spins and the colors, and a wrong prediction was then attributed to arbitrariness or caprice. In the course of the present stage, however, the independence of the separate trials and the structure of the whole due to the random mixture of the trials constitute the two new characteristics whose unification permits the construction of the notion of chance.

A third characteristic of these initial intuitions is added with Desp: This is the essential idea that regularity, which is due to the random mixture or the fortuitous distribution, increases with the number of trials *"because we can often hit another color but not often the same one."* We find this notion in Ank in an interesting form: It is more unusual to have one color untouched in twenty than in ten trials, and the reason for this, again, is because the mixture increases with the number of spins. For this reason the intuition of frequencies or distribution and the intuition of random mixture are really the same, and by this very fact lead to the beginnings of quantification. It is these three characteristics together which distinguish this stage from the preceding one.

On the contrary, if we compare this stage with stage III, we note the limits of this quantification. It does not, in fact, go beyond small groups, say of ten to thirty successive trials, and does not yield any generalizations for larger numbers. But we have here a paradox which we must examine. To say that the mixture of colors will be better with twenty trials than with ten, or with many trials—twenty or thirty as Chen ingeniously said—is to be already on the road toward the law of large numbers because this law states precisely that there will be an increase in the compensation with an increase in the trials. But a strange fact also to be considered is that at this stage this law holds true in the minds of the children only for relatively small numbers, and reverses itself when we go beyond about one hundred. Thus Tor imagines that the differences will be greater for sixteen hundred trials than for sixteen (*"five hundred, three hundred, two hundred make more"* than five, three, or two), and Web believes, in the same way, that we can have regularity *"more easily with fewer trials."* Derb and Flum also claim *"that it is more regular when we do it less often"* (*"I can't explain it,"* Flum says, *"but I understand it"*). Far and Chen reason in the same way: *"When there is chance, it's easier"* to reach uniformity with sixteen tries rather than with one hundred.

To what then can we attribute this extraordinary reversal in subjects who admit unanimously that mixture and, consequently, regularity of distributions increase with the number of trials? The children give two reasons and they are both instructive. The first comes from the failure to understand proportionality which we examined in Section 5 of Chapter II: To grasp the law of large numbers, it is important first that one recognizes that a difference of five in one hundred for example is less a difference than five in ten, since the differences act in relative and not absolute ways. Now we see that Tor calculates correctly when he says that differences of five, three or two out of sixteen would give differences of five hundred, three hundred, and two hundred in sixteen hundred trials, and that he concludes, therefore, that *"the dif-*

ferences are greater," as if their absolute values only were being considered. In the same way we see that Flum declares, still making this comparison of sixteen with sixteen hundred, that *"it would be the same thing. All we do is add zeros."* And he next concludes that *"this would be less regular."* But a second reason, a more important one, must be added to this first one. The law of large numbers does not, in fact, affirm the conservation of relative differences because it holds that relative differences diminish with the number of trials precisely because of the growing importance of compensation. The child of this stage will come to an understanding of this mechanism up to ten, twenty, or thirty trials because these are groups which can be observed concretely. But for one hundred, eight hundred and sixteen hundred trials, he is no longer certain of what can happen because he lacks the ability to see the number of possible representations or directions. This is the meaning of Chen's words which we cited, *"When there's chance, it's easier"* to have an equal number of stops per color with sixteen trials than with eight hundred because then the compensations due to mixing are exactly more easily imagined concretely for a small number than for a large one.

But in doing this, we simply push the question further back: Why are they able to generalize a process for the small large numbers, if we can say it this way, while they do not generalize for truly large numbers? The reason is tied to all the characteristics which we know about this stage. Proceeding by concrete operations, the subject does not succeed in reasoning except with data which are able to be handled either in thought or in fact, that is, visible or representable in detail. This is why he does not succeed in going beyond groups of twenty-five to thirty trials, those which are close in size to what he has just analyzed during the first trials. To apply the same scheme to large numbers he would have to have formal thought capable of deducing all the combinations according to a formal scheme operative to the second degree, that is, no longer concrete but operating on concrete operations by means of abstract thought. This formal or operative thought of the second degree appears only at about eleven to twelve years, in stage III, as we will see in Section 5 as well as in the third part of this work (Chapters VII through IX).

5. *The second stage: II. Reactions to the magnet.*

In the same way that they succeed in constructing relationships of variability to characterize the mixture, the subjects of this second stage seek also to construct a constant connection, without contradictions, which will take into account the stopping of the disc at the magnetized boxes. There are two points especially to be noted because they concern the use of the boxes, some of which contain magnets. First, there is the reaction of the subjects from the point of view of chance. In the course of the present stage, all the children, without exception, consider the systematic stopping of the bar in front of the magnetized boxes as caused by some factor other than that which the free spinning of the disc offers. We remember how, for the children of the first stage, both the phenomena were of the same order, or were more nearly so than is the case for those of stage II. In the second case, it is important to analyze the inductive reasoning itself which leads to the discovery of the causal relationship, and to show its connection, as we did before (Section 3) with probabilistic reasoning. Inductive reasoning, which begins to develop in the course of the present stage, consists in excluding certain factors and keeping others in such a way as to reconcile the data until we are able to draw from them a general coherent relationship. We see then both the similarity and the opposition between such logical steps and those which lead to the idea of fortuitous uniform distribution. While this distribution implies the notion of equal possibilities, that is, the inclusion of the multiple possible cases, the inductive discovery of a causal relationship supposes, on the contrary, the exclusion of certain factors (color, weight, etc.) and the use of others. It is important, therefore, to put these contrary but related concepts together; to do this we need to analyze briefly the operative procedures of exclusion and disjunction which the child at this level has. This is even more worthy of note because (as we will see later) the notions of probability themselves, in opposition to the notion of chance, and notably the relationship between favorable and possible cases, will include in their turn the intervention of disjunction.

The growth of inductive reasoning which characterizes this level is especially instructive about the operations of disjunction or of exclusion. The subject who is faced with the constant stopping of the bar at the magnetized boxes is quickly able to exclude the factors such as color or speed of spin, but they need to exert a much greater intellectual effort to exclude the factor of weight. We must remember that the magnetized boxes each had the same weight (B) as the boxes containing a nonmagnetic metal. These boxes weighed more than the boxes (C and D) which contained wax or powdered metal. By comparing the magnetized boxes with the lighter ones (C and D), the subject will be tempted to say that weight is the factor that stops the bar, but it seems to us that he simply needs to compare the weight (B) of the boxes with magnets to the heavier boxes (A) which contain lead, and especially notice that the non-magnetized boxes are the same weight (B) as the ones with magnets. This recognition should allow him to exclude weight as a determining factor. The subject at this stage, however, is still far from being able to organize by himself the control experiments which would allow him to eliminate weight as a consideration. This question concerns the psychology of experimental induction in general, however, and not the very structure of reasoning.[1] On the other hand, in considering this question of structure, we observe what happens when we encourage the child to make control experiments or when we put counterexamples in front of him, or results contradictory to his expectations. The youngest subjects exclude for a moment weight as a factor, but quickly come back to their first idea. Or they notice (to distinguish them clearly from those of stage I) that the stopping does not depend on weight as a simple function and, therefore, they try to reconcile the results with a coherent whole. For example, they will compare a median weight (B) with a heavy (A) or light (C) weight, or they will relate one of these weights with the weight of the disc, and so on. Only those subjects nine years old or above realize weight is not a factor after they have seen counterexamples which we have constructed. We shall try to determine why this exclusion of weight is so much more difficult for them than exclusions of color or speed:

DAN (6; 8): Magnetized boxes on white: *"It moved all by*

[1] At the present time Bärbel Inhelder is making a study with us of this problem which goes far beyond the special question of chance.

itself!" "Try it again." *"Look! It's the box of matches that makes it move."* (We remove the boxes leaving only the nonmagnetized ones on black.) "Where will it stop?" *"On black. I don't even have to look. I know it."* (Stops on red.) "What's the matter?" (He grabs the boxes.) *"They are light* (weight D), *they are nothing at all. They have to be heavy."* (We give him the heavy ones, [A], but without magnets.) *"Oh, these are very magic because they are very heavy."* (Experiment.) *"No* (with surprise)." "Now what?" (He weighs the boxes.) *"It's the boxes which weigh more that stop the disc. These* (C and D) *don't weigh much. I don't believe that they are heavy enough, otherwise the bar would stop. These are too heavy. It's the middleweight one that works."* (We then have him compare the boxes of weight B, those with and those without magnets.) *"They are the same weight."* "O.K., then is it the weight that makes the bar stop?" *"No."* But he later comes back to his hypothesis that the boxes of median weight work.

DESP (7; 6): *"It's because the boxes are there. They hold, they're close. When it turns fast, it goes by, but when it goes slowly, it can't pass. It's tight, it weighs more, it makes it lean."* (We leave only the light boxes.) *"They aren't heavy enough to stop it."* (We leave only the A boxes.) *"No, yes. Maybe they are too heavy. It can't very well stop like that. They weigh too much. When there's a little less weight, it weighs just a little, and it can make it lean better."* (He distinguishes between "weigh" and "lean" by leaning on the edge of the disc.) "Then why doesn't it work with the lighter ones?" *"They don't have enough weight to make it go."* "And these (with magnets)?" *"That'll work, that's the right weight."*

TOR (8; 4): *"I don't know why the boxes make it stop. Is it because they weigh like that?* (We remove all but the light ones.) *"No, that's not heavy enough.* (Boxes A put in place.) *That will do it."* (Experiment.) *"No, it was too heavy."* "What, then?" *"They are too heavy or too light. Those* (with the magnets) *are just the right weight, and it'll work with them."*

FUL (8; 4): We place the boxes and Ful notes that the bar stops on red (at the magnets). "And now where will it stop?" *"We don't know."* (Experiment.) *"Again at the same place."* (We do it again.) *"Again the same thing. There's a mechanism inside."* (We change the positions of all the boxes.) "Where will it stop now?" *"Same place. You changed the boxes, but it will still stop on red."* (Experiment.) *"Oh, no. It stopped on green. It always stops in front of those boxes.* (He opens them). *"There's something inside, wax. Is there some in all the boxes?"* (He checks). *"Yes, but this one is lighter, that one heavier. It could stop before all of them. I don't know why, but it's always in front of that one."* (We change the positions of the boxes, the magnetized ones on yellow.) *"It will always stop in front of those. The color doesn't matter."* "What now?" (He reweighs them.) *"This one* (A) *is the heaviest. It's only the middleweight ones that stop the bar. I don't know. It's difficult."* "Are you sure that these (the magnetized ones) are the only middleweight ones?" (Ful compares the boxes [B] containing the magnets and those which have the nonmagnetic metal.) *"It's the same weight."* "Then does it depend on the weight?" *"Some are heavier than these* (magnetized ones)."

Duf (9; 4): "If I spin it many times, where will it stop?" *"On red, yellow, on all the colors."* (We make several trials, then we place the boxes, and five times in a row it stops on red where we placed the magnetized boxes.) "Where the next time?" *"It'll stop ten times on red and then on another color."* "Is it usual for it to stop ten times on red?" *"Yes, up to ten, it's normal, but more than that is not normal."* (We continue.) *"No, it's not usual. It's because we're not turning it fast enough."* (He spins it faster but it stops again on red.) "Why? Is it because of the colors?" *"No, it's because of the weight of the spinner* (disc). *It's too heavy."* "But why does it stop on red and not on blue or green? . . . It is because of the boxes?" *"No."* "How can we be sure?" *"Take away the boxes."* (He takes them away.) "Where will the disc stop this time if your idea is right?" *"On yellow."* (He spins; the bar stops on yellow!) "What does that prove? That your idea was right?" *"No, it's because of the boxes. These* (with the magnets) *are heavier than the others. It never stops on blue or green because the boxes are too heavy* (boxes A and B without magnets)." "How should we do it? . . . Take them away?" *"Yes."* (He does the experiment without the heaviest boxes and then without the lightest ones.) *"It is the weight of those two boxes* (B, with magnets) *because they are neither too heavy nor too light.*" (We have him compare the weight of the boxes of weight B, those with and those without the magnets.) "*It is the same.*" "Then is the weight the reason?" "*No.*" "O.K., if it is neither weight nor color nor speed of the spinner, then?" *It's because of the wings* (the brakes of the disc)." "What do we do?" "*Change their place* (he moves them and that changes nothing)." "Now?" "*It's what's in the boxes.*"

DEN (9; 1), in the same way, compares the weights and says, "*There are some heavier than others* (B greater than C; B greater than D). *These* (A) *are even heavier.*" "And where does the disc stop?" "*It stops on those* (B). *Right, it's not the weight.*"

FAU (9; 5) in the same way: "*It's not the weight because these* (A) *are the heaviest and the spinner stops on a lighter one.*" "Give me another reason why weight makes no difference." "*It could just as well have stopped at those* (B without magnets) *as at those* (B with magnets)."

We see the interesting reasons for comparing these reactions to those of the same subjects when there was simply random stopping because we were not using the magnets. Contrary to the subjects of levels I A and I B, each of these subjects shows, in fact, immediately the feeling that the regular stopping of the bar at a certain point could not be fortuitous and depends, therefore, on new causes. Dan, for example, notes from the beginning the causal action of the boxes; Duf sees the bar stopping constantly on red and decides for himself that if simple chance were working the repetition could not go beyond the number ten because "*up to ten, it's normal; more than that is not normal.*" In other words, since without the boxes the bar stops anywhere, and with them the bar stops always at the same place, they know that in the first case there is only chance working while in the second there is a particular cause at work.

The simultaneous apprehension of these two contrary situations depends on the same reasons and the action of the same mental operations. On the one hand, there is certainly the notion of uniform distribution, thus of the equal possibility of stopping on any of the different sections, while in the other case there is centered distribution on a single point, thus a unique determination. But to understand this idea of equal possibilities, the possible cases must be separated into a certain number of mutually exclusive cases (corresponding to the eight sections of the circle) seen as equivalents and with a balancing of the causes so that none are excluded. To understand why the distribution is centered on a single point is, on the contrary, to postulate the interaction of certain determined causes to the exclusion of others, that is, the separating of the mutually exclusive factors according to an either-or process which consists of deciding whether or not it has an effect. In both cases, thus, the understanding of the distribution supposes the same operative structure. This structure is absent at stage I, and its development explains the recognition of chance and the beginnings of deductive reasoning: It is the ability to include or exclude, that is, make an either-or choice among equivalent cases.

This new operative attitude is indicated first in the activity of the subject who, instead of remaining passive as in stage I, tries to understand what is happening by exploring all the apparatus. Thus Ful immediately opens one of the boxes in front of which the bar stopped and noting the presence of wax, wonders if they all contain wax and proceeds to verify this. All the subjects weigh the different boxes and compare the distinct weights. Desp differentiates the actions of leaning and weighing by his actions. Others compare the smooth or rough surface of the wax in the boxes or mention the action of the weight of the disc. Duf, especially active, varies the speed of the spin, removes and replaces the boxes, eliminates the heaviest and the lightest, shifts the brakes of the disc, and so forth.

But most importantly, and certainly a part of this meaningful experimental activity, the subject begins to exclude certain factors, an operation quite beyond the children of stage I (who tried to reconcile everything, even when it meant contradicting themselves, because they were unable to comprehend exclusion). It is for this reason that at this stage the factor of color was eliminated immediately, as soon as the subject perceived that the fixed stop was not tied to any determined section of the circle. In this way, although Ful was tempted to believe that the color red played a role, he noted, *"Oh, no, it stopped on green. It always stops in front of those boxes . . . it has nothing to do with color."* The speed of the spin was also easy to eliminate: Duf supposed that the regular stopping of the bar was due to the force of the spin and tried to spin the disc faster, and concluded immediately, *"No, it's not that."* In the same way, in trying to decide if the cause is to be found in the boxes or outside of them (dichotomic disjunction), he removed all the boxes and then noting that the action stopped (*tollitur effectus,* the cause removed), he eliminated external causes.

As for the weight factor which all the subjects thought was a cause when they noted the differences in weight, this was little by little eliminated, but with more difficulty than the preceding factors had been. The youngest subjects of this stage (seven to eight years old) still have a strong tendency to try to reconcile everything, but they differ from children of stage I in that they avoid contradictions: If the heaviest boxes (A) do not stop the disc and the lightest ones do not either, then it must be a question of an optimum weight to achieve the desired effect; and this optimum weight will be the median one (see Dan, Desp, and Tor). This is a curious hypothesis, and the next steps in the development show that the more advanced subjects conclude that the weight factor must be excluded after noting the boxes of weight A produce no effect. However complicated this may be, it is not an absurd idea. The situation is spoiled when the child is forced to compare the boxes of weight B which stop the disc and those of the same weight which do not. (He has a great difficulty in doing this just by himself.) Dan decides at this point to eliminate the weight factor, but nonetheless, he next comes back to the hypothesis of an effective median weight. With children from about nine years of age on, however, we witness clear exclusions of the weight factor: Ful and Duf, although tempted by the solution of choosing the median weight, definitely cast it aside when they compare the boxes of weight B. Den decides against the factor of weight after noting that the heaviest boxes (A) do not stop the disc: *"O.K. then, it's not the weight."* Fau, finally, uses two instances of experimental evidence taken from the boxes of weight A and those of weight B without magnets.

Thus weight itself is little by little eliminated by the subjects of this stage. We can see quite well why this exclusion takes more time than the exclusion of color or speed because, on the one hand, it is a question of a quality inherent in the boxes themselves and not in exterior factors; and on the other hand, once the child has eliminated weight, there is no other intelligible cause for him, except to say as Duf did, *"It's what's in the boxes."* It is clearly easier to apply the operation of reciprocal exclusion to a pair of choices such as color or not color or speed or not speed than when we know that the elimination of one factor still leaves several factors to choose from.[1]

Leaving aside this relatively long time needed to solve the problem of weight (especially interesting for the problem of induction in general), we note that the subjects of stage II are likely from the beginning to make certain exclusions and, as a result of this, to seek causes for both controlled and nonuniform distribution at the very moment when they are also becoming capable of understanding the mechanism of a uniform fortuitous distribution. Such a relationship between the construction of two such antithetical notions is to be carefully noted and will be very useful for us later when we pass from the study of chance to a study of the quantification of probabilities.

6. *Third stage: The beginnings of formal reasoning in the discovery of the relationships of variability and constancy.*

In considering the fortuitous distribution (before using the magnets), the progress shown in this stage is the discovery of the law of large numbers. As for the inductive reasoning of the constant relationship, it is characterized by two new elements: a greater ease in the improvisation of experimental methods which permits the subject to dissociate the factors at work; and, at the same time, an ability to reason in a hypothetical-deductive manner, that is, gen-

[1] No doubt there is here some of the lack of differentiation proper to the notion of weight itelf which is conceived of traditionally as the expression of force in all the meanings of the term (and here as a force of attraction).

eralizing and excluding by means of thought as a function of a set of implications, and no longer only from concrete observations.

Here are some examples of judgments of probabilities, beginning with two cases intermediate between stage II and stage III:

BAL (10; 7) refuses to guess on what colors the bar will stop *"because it moves, it can stop at any color."* "But could there be someone who might know?" *"Oh, no, unless, of course, there were something that stopped the bar."* (He says this before the use of the magnets.) (Experiment: brown.) "And now will it land again on brown?" *"It's less sure to land on brown because there are so many other colors."* "Will it hit all the colors or not?" *"It depends on how long we spin it."* "Why?" *"Because if we spin it often, it will have more chances of going everywhere."* (Cf. the law of large numbers!) A moment later: "We have spun it sixteen times. If we do it sixteen hundred times, will it be more or less regular?" *"It will be more regular."* "Why?" (Doubt.) *"Oh, no, less regular because with sixteen there's more chance of it falling equally on each color since there are fewer spins."* "Why?" *"When there are fewer, it's more regular because there are fewer gaps."* "And with sixteen hundred?" *"Oh, it will be the same."*

JAC (10; 8): Same initial reactions: *"It's chance, it's something that can't be predicted."* After fifteen trials: "If I did it three hundred times, would the differences be relatively larger or smaller?" *"The same* (but after reflection), *no, larger."* "If we did it eight hundred times, would it be more likely to stop on certain colors rather than on others, or would that be more likely doing it only fifteen times?" *"With fifteen the inequality is more likely than with eight hundred. When we spin eight hundred times, there is more chance and it goes more to all the colors."* He comes then finally to the correct answer by using the word *chance* in the sense of random mixture and compensation.

MIC (12; 1): "If I spin it eight hundred times, will it be more or less regular than doing it fifteen times? *"More regular."* "Why?" *"It will equalize."*

LAU (13; 4): "If I spin it eight hundred times will it be more or less regular than doing it fifteen times? Will there be the same number of stops on each color, or will it be more regular with fifteen?" *"It will be more regular with eight hundred because that's larger. The more times we try it, the more chances there will be.*" "Why?" "*Because of the more trials.*" "But why is it more regular with more trials?" "*I don't know. Because there is more chance.*" "But why is it more regular when there is more chance?" "*O.K., for example, out of one hundred trials, there will be a certain number of stops on each color, but not the same number. For two hundred trials, it will become more equal, and the more we do it, the more equal it will become.*" "But why does it become equal?" "*I don't know. Yes, exactly, there are more chances because there are more trials. For a small number, it changes each time and it depends on the number of times, but with a large number of trials, it has more chances of being more regular.*" "I don't understand why." "*When we do it just a few times, once it goes and the next time it doesn't. But when we do it a large number of times, then it becomes regular because it stops one time here and once there, and after a certain number of tries, they are all the same.*"

The meaning of these few cases could not be more clear: The understanding of the law of large numbers is derived from the intuition of a progressive random mixture, an intuition whose appearance characterized stage II. The understanding of the law begins when the intuition is generalized thanks to formal operations beyond the limits of concrete observable groups. Thus Bal has already formulated the law of large numbers when he says, "*If we spin it often, we have many more chances of hitting everywhere,*" that is, of achieving regularity. When we suggest to him sixteen hundred tries (which goes far beyond what he had thought of as often), he is then brought to make the generalization, "*That will be more regular.*" His idea, however, does not follow the direction indicated by this deduction, and he backs up, first using a mode of reasoning common to stage II (fewer trials means less of the unexpected), and finally concludes, "*the same thing.*" As for Jac, he has the same initial reactions, but finally realizes that "*when we spin the eight hundred times there is more chance,*" which gives the sense of a greater regularity by compensation in the fortuitous differences. (This answer of Lau is the contrary of what Chen said, stage II: "*There is more chance in sixteen than in eight hundred.*") Finally, Lau formulates in the most explicit way the compensating mechanism proper to the law of large numbers. And the curious part is that he does it in almost the same terms used by the subjects of stage II (Desp, for example, seven years, six months), when they are explaining small groups of elements. It is obvious, therefore, that what distinguishes the subjects of stage III from those who are limited to the law of small large numbers is the difference between a physically perceptible situation and an abstract generalization along with the understanding of the proportionality of differences in number. This generalization which goes beyond the concrete and this proportionality demand, in fact, formal thought, the former because it implies an extension of all the possible cases of the combinatoric operations, and the latter because it demands the construction of links between the relationships.

As for the induction of the constant relationship with the introduction of the magnets, following are a few of the typical cases:

GUG (12; 7), at the third trial after the boxes were put in place, is surprised that *"it always stops at the same place,"* and decides to spin the disc himself *"to see if it will stop again at the same place."* (Experiment.) *"Yes, it will always stop there."* (He lifts the boxes.) "Why did you do that?" *"To see the weight. The weights of the pairs are the same."* "So?" (He removes one of the magnetized boxes and notes that the stopping is less clear.) *"It's the distance."* (He moves the other magnetized box further off, and the bar stops elsewhere.) *"Once we take away one of the weights, if we take away the other, we get the same result."* (He puts the magnetized boxes back on the board and takes away the boxes of weight B without magnets which he has found out are of the same weight as the first ones.) "Why are you doing that?" *"To see if it stops at these* (magnetized) *or those* (not magnetized)." (By doing the experiment, he finds out that only the first ones attract the bar. Then he moves all the boxes out about an inch except the lightest ones, D.) "Why?" *"To see if it will stop at the closest ones."* (Experiment.) *"No, those which are further away have more influence than the closer ones."* (Then he moves the other ones in closer and tries.) *"The closer they are, the more they attract."* "So?" (He moves away all but the heaviest ones, A.) *"It's not the weight that works it because it doesn't stop at the heaviest ones which are the only close ones.*" "What does that prove?" *"There is a magnet attracting it."* (Again he tries with the magnetized boxes and the boxes of the same weight without magnets.) *"Since they are of the same weight, weight is of no importance because these always make the bar stop."* "If I move these away a little (magnets), and it stops at these (the heaviest ones), what will that prove?" *"That proves that the magnet is not strong enough if it is too far away."*

ORA (13; 4): At the second trial after the boxes have been put in place: *"It must be magnetized; it even pulls it back; there must be something in the boxes."* (He examines the boxes.) *"These* (B) *are heavier than those* (D). *They are of different weights."* (He compares all the weights with the weight of the magnetized boxes.) *"It must act on the spinning disc."* "What can we do to prove it?" *"We could put them in different places."* (He notes that the weight A does not stop the bar and then, while moving a magnetized box, notices that the bar moves.) *"Watch, it follows the box."* (He tries the same thing with several different boxes, and then takes away all of them except the ones without magnets.) "Why are you doing that?" *"To see if these will stop the bar, and so as to get no interference from the other boxes."* (Experiment.) *"They have no importance."* "Does the weight have any influence?" *"No. These* (A) *which are even heavier don't attract."*

KRA (13; 11), who doesn't think about the possibility of magnets, nonetheless excludes weight as a factor by comparing the boxes. *"In any case, it is not the heaviest ones* (A) *which do the attracting. And these* (B with the magnets) *are the same weight as those* (B without magnets), *and only these* (the first ones) *attract."* "And so?" (He moves the ones with magnets somewhat away.) *"I want to see if they have the same influence at a distance that they did close up. . . . I think it's rather the substance of what's inside."*

The difference between these subjects and those of stage II is seen in two related elements of their activity. First their experimental activity is clearly superior: They themselves dissociate factors, organize counterproofs, and succeed in eliminating or establishing inductively the role of certain variables such as weight or distance.

Whether they guess or not that there are magnets hidden in the boxes, they proceed along the same methodical steps, varying the data and especially by modifying a single relationship while holding aside the group of other factors as a means of verification. This greater flexibiilty in transforming the experimental situation results in a second new element proper to this stage: the ability to reconstruct reality by a hypothetical-deductive method. It is this formal thought whose role we just noted in the discovery of the law of large numbers, which also explains the perfecting of the operations of exclusion. Already present in stage II, exclusion is seen here simply in the disjunction between concrete classes without common elements (color or not color, weight or not weight). On the formal level which concerns implications between propositions and no longer only concrete relationships of classes or linkages, exclusion becomes an interpropositional operation which is immediately applied to the play of hypotheses: If weight were the cause, says one subject for example, the heaviest of the boxes ought to be most effective. Since this was not the case, weight is definitely excluded.

We see, therefore, in the total picture, the close relationship between the appearance of inductive mechanisms and the culmination of the notion of fortuitous distribution. And in the same way that the discovery of equal possibilities went right along with the discovery of concrete exclusions in level II, the understanding of the law of large numbers appeared in level III in correlation with the formal procedures which made possible the first systematic inductions and the first deductively constructed exclusions.

PART TWO

Random Drawings

IV. Chance and "Miracle" in the Game of Heads and Tails

After studying the development of certain intuitions of chance as a function of physical phenomena, it is essential to analyze a second aspect of the notion of the fortuitous: that which is developed in games called games of chance, which are well known to both adults and children and which, in the history of science, have brought about a theory of chance and calculation far ahead of investigations of physical probabilities. No doubt dice games, heads and tails, taking counters from a jar, and so forth, have a physical aspect and, therefore, are related in a sense to physical probability. But there is an interaction in these games quite a bit more complex when compared to the experiments of Chapters I through III. This consists of actions or operations of the player himself who is no longer limited to observing from outside a mixture or a distribution, but who draws a lot, bets, and so on. It is these operations, or activities, of the subject which give to games of chance a different psychological tonality than is to be found in the mere observation of physical chance, and which are the starting point of the application of logic and mathematics to chance. This is why we are devoting the second part of this study to an analysis of the reactions of the child to playing heads and tails, to drawing marbles from a jar, and to the quantification of probabilities in random drawings.

The present chapter will take up the game of heads and tails. As in Chapter III, we have tried to compare the intuition of fortuitous distributions with the discovery of a constant causal relationship. To get a better idea of the notions of chance that the child has in the game of heads and tails, we will try to contrast the free distribution of throws in the normal game with the "miracle" in which a run of only heads or only tails occurs. Chance is, in fact, the negation of miracle—that is, to understand the nature of an uncertain distribution will mean for the child, as for us, admitting the very weak probability, or even the practical impossibility, of an exclusive succession of heads or tails. Thus we have furnished our subjects with an apparatus which will allow us to produce the miracle at will, to cheat chance, if we can say it this way, so as to analyze the reactions. Will they still have the impression that everything is normal when they are faced with this astonishing fact; and what will this impression mean from the point of view of the notions of frequency and rarity? Or will they readily get the impression of impossibility, in the sense of an extremely small probability, which will be manifest in the immediate suspicion of an intervening trick (the equivalent of the intuition of a nonfortuitous causal relationship as was the case with the magnets in Chapter III)?

For this purpose we have used two distinct tests. The first one consists of presenting white counters marked with a cross on one side and a circle on the other. We play heads and tails by having the subjects predict the outcome when ten to twenty counters are thrown at once. Then, without letting the child suspect any substitution, we throw about fifteen fixed counters having the cross marked on both sides. And we repeat this experiment until we get a sufficient understanding of the reactions of the child. The other apparatus consists of a sack with red and blue marbles inside. After having the child predict and note results of the experiments, we substitute for this sack another which has only blue marbles. In both tests, after the subject's reactions have been analyzed, we finally show him the trick, if he has not already discovered it, and then we do a last experiment: Without his knowing whether we are using the fixed set or not, we throw one by one the trick counters or take out the marbles one by one from the trick sack (blue marbles). Then we try to catch at what moment and thanks to what reasoning the subject recognizes for sure that it is an experiment with the homogeneous, or fixed, elements. This last experiment often gives the surest index of the judgment of probabilities of which the child is capable.

We can distinguish three stages in the reactions to the games of heads and tails and to the test with the fixed counters. In the course of stage I (up to about seven years of age) the child is more or less surprised (although sometimes not at all) by the fact that they came out all heads, but does not succeed in understanding its impossibility from the point of view of probability. Even after the discovery of the trick (pointed out at different stages of the experiment), he thinks in general that with normal counters the same miracle could happen. The second stage (seven to eleven years) is characterized, on the contrary, by a global sense of the probabilities, but the final choice between the two hypotheses (normal or fixed group of elements) does not yet show a nuanced quantification of the combinations. This appears finally only in stage III. The test with the marbles is a bit simpler.

1. *The first stage: Intuition of rarity, but not of random mixture.*

Here are the facts first:

Knu (4; 7): "If I toss this counter, will we see the circle or the cross?" *"The circle."* "Sure of it?" *"Yes. Maybe the cross, I don't know. I'll see."* "Toss it yourself." *"A circle."* "And this time what will it be?" *"Again a circle."* "Try it." *"A cross."* (He does it again.) *"Again a cross."* "What if we toss the whole pile at the same time?" *"Some will land with the cross up, and others with the circle."* (Experiment.) *"One more of the circles."* "Once again?" *"They will again land with some of both up."* (False counters.) *"All crosses. That's odd."* "How did that happen?" *"Like that."* "But how? . . . Is there a trick?" *"Yes, a trick that you can do with your hands."* (We do it over again shaking them hard.) "What is happening?" *"I don't know. That's odd."* "Try it yourself." (He empties the sack.) "Why all crosses?" *"I don't know."* "Do you guess the trick?" *"I don't know. They all fall the same way, same side."*

Pac (5; 0): "Which side will be up?" *"Don't know."* (A circle.) *"And now?" "A cross."* "Why?" *"Because."* "If we throw a lot of them, will they all turn up crosses?" *"No, circles and crosses."* "Many circles and few crosses?" *"No, many circles and many crosses."* "Very good. Why? . . ." (False counters.) *"Ah, all crosses."* (He is surprised and pleased.) "Why?" *"Don't know."* (We try again.) *"Still all crosses."* "How did it happen?" *"Don't know."* "Is there a trick?" *"No."* "Can it happen that they all land the same side up many times?" *"Yes, because we are throwing a lot."* "But I say that there is a small trick to it. Think a minute." *"Don't know."* "Look." (We turn over a counter.) *"A cross on both sides."* "And with these counters (we take the first sack, with circles and crosses, and show him both sides of the counter), can it happen that they would all turn up crosses?" *"Yes."*

Wi (5; 2) seems to have had no experience with the game of heads and tails. "If I throw this counter will it fall with the cross or the circle up?" *"I don't know. If someone has never seen it, he can't know."* "What should we do?" *"We have to try."* (We toss the twenty counters one by one; Wi becomes more and more interested in the game.) *"Oh, a little cross! And a little circle!"* (At the end there are more circles than crosses and Wi makes two groups to compare the results.) *"There aren't enough crosses."* "Why are there more circles?" *"Don't know."* "What if we threw them again?" *"Maybe some of both sides."* "How many of each?" *"We'll see. Maybe more crosses because last time, there were more circles."* "Could they all be circles?" *"Maybe one cross and all the rest circles."* (Experiment.) *"Mixed. . . ."* (False counters.) *"All of them are crosses."* (He shows no surprise.) "How does that happen?" *"Don't know."* (In playing with a counter, Wi finds a cross on the other side.) *"Oh, a cross on the other side too."* "All right, and if we do it again, how will they fall?" *"Don't know."* All crosses or all circles?" *"Don't know."* "Or half and half?" *"Yes."* (Experiment.) *"Oh, all crosses again."* (He turns one over.) *"On the other side too."* "Do you think that there is a cross on the other side of every one?" *"Don't know. We have to look at all of them."* (He does it.) "And with these true counters, which have a cross on one side and a circle on the other, can it happen that all land with the cross up or the circle up?" *"Yes."* "If we toss a lot of them, will they all fall with the cross up?" *"Don't know."*

CHAP (5; 4) thinks that the first counter will fall with the cross up *"because it's larger* (than the circle." "Try it." (He makes a toss.) *"It's the circle."* "And next?" (Experiment.) *"Cross."* "Now if we throw all of them together, could all of them turn up crosses?" *"No, they couldn't all fall crosses because they turn and hit each other as they fall."* "Why?" *"That's the way it works; they can't fall all crosses or all circles."* (False counters.) *"All crosses at the same time!"* (He is amazed.) "Try it yourself." *"Yes, they all fall crosses up."* "Do it again." *"Again all crosses."* "Again." *"That's funny, again."* "Why is it like that?" *"They all fell to that side as they came down."* "Don't you think there might be a trick? . . . Look at it." (We give him a counter.) *"A cross on both sides."* "And with the true counters (we show him the sack and the counters with crosses and circles), could they all fall with the circle up?" *"Yes, if we want them to."*

MAR (6; 0) predicts a circle and then a cross during several trials, one at a time. "If we toss a whole pile of them, can they fall all with the same side up?" *"Yes, it could happen."* "What side?" *"Circle."* "Why?" *"The crosses would all be facing down."* "Would we also get half-crosses and half-circles?" *"Yes."* "Watch." (We throw twenty.) *"Yes."* (Just about half and half.) (False counters.) *"All crosses."* (We do it again.) *"Again all of them with little crosses."* "Why?" *"Because all the little circles are turned down."* "What if we do it again?" *"They will all be little circles."* "Let's try again." (We give the sack with the real counters to the child and we toss the false counters.) "Why are yours mixed and ours are all crosses?" *"Because the little circles are all turned down in yours."* "But why are they all turned down in mine and not in yours?" *"Don't know."* "You don't think there's a trick?" *"No, not if you threw them like this."* "O.K., throw them yourself." (We give him the sack of true ones.) *"There are crosses and circles."* (We give him the sack of false counters.) *"Nothing but crosses. That's because we threw them hard.*" "O.K., throw them so that they will be mixed. (He tries.) Why didn't you get some circles?" *"Don't know."* "Is there a trick?" *"No."* "But I say yes there is." *"It's because we threw them too hard."*

CLAI (6; 11), foresees a mixture *"because they don't all fall on the same side."* "What if I throw one thousand of them?" *"There will be three hundred crosses and seven hundred circles."* "Why?" *"Because they won't come out all the same thing."* "Let's try it with this sack." (False counters.) *"All crosses."* "Why?" *"Don't know."* "What will it be if I do it again?" *"Maybe one circle and the others crosses."* "Why?" *"Because there are already so many crosses."* (Experiment.) *"Again nothing but crosses."* "Is there a trick?" *"Oh, certainly not."* "I say there is one." *"There is none."* "If we tossed one thousand, could they all come out crosses?" *"Yes, that could happen."* "Often?" *"No."*

These are examples of the most elementary reactions we have observed in children more than four years old. A certain number of constant tendencies can be found here, and they characterize the first stage.

First, there is phenomenism, or a certain docility when subjects actually observe an experiment, a reaction quite different from the experiments used to test a previous hypothesis. This is quite a natural attitude for children, for as Wi says, *"if I've never seen it, I can't know."* It is, however, an attitude which can lead one to seeing as quite natural anything that happens and accepting appearance as well as reality. This is why none of these subjects thinks of a trick when faced with the false counters until we suggest it. From the point of view of phenomenism, there is no miracle when all the counters turn up crosses, but only a new fact.

A second characteristic, deriving from the first one, is what is called passive induction,[1] or empirical induction (as distinct from active or experimental induction), based on the simple intuition of what is frequent or rare. For the majority of these children, tossing the counters will give as many circles as crosses, but this

[1] Passive in a relative sense, that is, intermediary between transfer and generalization.

is simply what the experiment has taught them and not what they deduced from a combinatoric operation. This is why the false counters do not seem strange to them or difficult to accept without remark, but rather natural or simply funny, because rare (Knu). Wi who was seeing the game of heads and tails for the first time sees it as quite normal and is not even surprised at it, in spite of his predictions to the contrary which we will return to later. Chap reacts in the same way, finding immediately an explanation whose tautological character shows the purest phenomenism: *"They all fell to that side as they came down."* In brief, the reaction of those experienced with the game is that they are face to face with an uncommon experience, and for the others, a lack of understanding which leaves them surprised at nothing.

But if this is the phenomenist side of their reactions, there is another reaction quite as important and which we know from preceding chapters. We find this third characteristic quite clearly in Wi whose inexperience is in this case valuable: After the first five trials this child concluded suddenly, from the fact that the circles were more numerous than the crosses, that the next group would give *"more crosses because the last time there were more circles."* Thus there is a tendency to see compensation and equilibrium, a tendency in which factors of perceptive symmetry, justice, and sufficient reason no doubt play a part (there is no reason for one side winning out over the other). Thus we have numerous motives which are at the same time subjective and objective.

The fourth characteristic derives from this third one as the second did from the first. When the child seeks an explanation other than a subjective one (*"they fell that way"*) for the fact that all of the counters show crosses, he decides that it is due to the power of the person who threw the counters. *"It is a trick you can do with your hands,"* said Knu. This *"trick with the hands"* which Mar also brings up saying, *"You threw them like that,"* and which he thinks he can do also with no trouble consists simply in throwing the counters with a certain force. The child has noted the mixture and seen the counters hitting each other as they fall (*"They turn as they fall,"* said Chap to explain precisely during the first part of the experiment that they could not all land with the same side up), but they can still imagine that all one needs to do is to toss them in a certain way so as to get only crosses showing.

Following is a résumé of the reactions of the subjects of stage I: When the counters fall at random, they explain the result by an intuition which is both phenomenist and egocentric (as is every preoperative intuition). The phenomenism is seen in the notions of frequency and rarity, still empirically known, and the egocentrism is seen in the subjective aspect of the idea of compensation. When the counters all fall with crosses showing, however, their phenomenism allows the children to accept the fact without difficulty (*"that can happen"*) and their egocentrism is to be attributed to the act of throwing as a means of avoiding the mixture. In summary then, these subjects have no more an exact notion of random mixture than did the subjects of stage I whom we saw in Chapters I through III. Even though they can observe the material and empirical act of mixing, they do not perceive the results as a system of combinations and permutations which are logical and mathematical, and this is the reason why they lack the notions of chance and probability and seek instead hidden controls in the mixture itself.

The best confirmation of the merits of these interpretations is the reaction of the child to the true counters, those with both a cross and a circle, after some trials with the false ones. Each subject is, in fact, informed by us at the end of the session about the trick of which they were the amused victims. Nonetheless, when we go back to the true counters, they still imagine that they can also come out all crosses. Note first the very instructive reaction of Wi (the one who knew nothing about the game before). When he sees the false counters all turn up crosses, and he turns one over by chance and finds a cross on the other side, instead of being suddenly enlightened, he concludes nothing about the rest of the collection. It is quite normal for Wi later, with the true counters, to imagine that they also could all land with crosses up, even after he had seen that the second sack was fixed. Chap was of the same opinion. Clai was even able to admit the possibility for a thousand counters, recognizing that it would *"not happen often."* No doubt it can be said that a child who has just been fooled will later acquiesce to anything. We will see, on the contrary, that in stage II this is not so, and this is why the question is asked of the subjects of stage I.

2. *The second stage: Chance and total possibilities.*

The most striking characteristic of the reactions of stage II is the refusal to accept the results of the experiment with the false counters. The child does not believe that this could happen without some trick. On the average this reaction is observed after seven years, but we have some examples of it from six years.

HAN (5; 11, advanced): "How will they fall?" *"One side and the other."* (We toss a few.) "What if I continue up to two thousand?" *"There will be one thousand crosses and one thousand circles."* (We toss out the false ones.) *"All crosses!"* (He thinks for a moment.) *"It's because there are crosses on both sides."* "Could the true ones fall that way?" *"No, because they fall some one side, some the other. They could never all fall with the same side up."* "Why?" *"They can't fall all with the same side up because they turn as they fall, by chance."*

BRA (6; 5) foresees that sometimes there will be more crosses than circles and other times more circles. We toss the false counter. *"Nothing but crosses!"* "What if I do it again?" *"Same."* "Why?" *"Maybe because there's a cross on the back too."*

BER (7; 2): When we throw the counters one by one, he foresees at the second toss, *"probably a cross."* "Why?" *"Because before there was a cross."* (The counter falls again a cross up.) "And now?" *"Circle."* "Why?" *"It can't always fall on the same side."* "Why?" *"Don't know."* (But he continues to predict by pairs.) "What if I throw them all at once?" *"Both sides."* "Both equal?" *"No, many of one and few of the other."* (Experiment.) *"About the same number of each."* "Is there a reason for that?" *"No."* (False counters.) *"Oh, surely they are the same on both sides*" (his immediate reaction without looking). "Could the true counters have fallen all with the same side up?" *"No, there are too many and they are too mixed."*

Mal (8; 7) refuses to predict in detail. *"They fall with any side up."* "What if we throw all of them?" *"Crosses and circles."* "More of one, or about the same of each?" *"No way of knowing."* (Experiment.) *"About the same."* "Why?" *"Because they are mixed."* (False counters.) *"All crosses!"* "Why?" (Two more trials.) *"It's because there are crosses on both sides."* "How did you guess? Did you know it on the first throw?" *"No."* "On the second?" *"Yes, but not absolutely sure."* "The third?" *"Yes, absolutely."* "Could we get such a result with the real ones?" *"No."* "What if I did it one thousand times in a row?" *"Maybe."* "In ten thousand times?" *"Maybe."* "More likely in one thousand or in ten thousand trials?" *"With ten thousand."* "Why?" We tell him that we are going to take counters from one of the sacks one by one and toss them to see if he can guess from which sack we are taking them. "First one." *"It's a counter with two crosses."* "Certain?" *"Not certain."* "Second one." *"Can't be sure."* "Third." *"It's the counter with two crosses."* "Sure?" *"Yes."* "It couldn't be a real one and fall three times in a row cross up?" *"Yes, it could."* "Four times?" *"Yes."* "Five times?" *"Yes."* "How many times are needed before you know for sure that there are crosses on both sides?" *"Don't know."*

Man (9; 5): "What if I throw them all?" *"They'll fall all over, some with crosses and some circles."* "Is it possible for them to fall with all crosses or all circles up?" *"Yes, but not often."* "Out of one thousand times, how often will it happen?" *"Two or three times."* "Out of ten times?" *"Once, I think."* (False counters.) *"They're all the same."* "Why?" *"Maybe there are crosses on the other side."* "Are you sure?" *"No, maybe* (he looks). *Yes, there is one on the back."* "And with the true ones, could it happen?" *"Possible but rare."* We ask Man to guess if the counters we are going to draw out come from the sack of false counters or true ones. "First." *"Maybe there's a cross on the other side, maybe a circle."* "Second." *"Same."* "Third." *"Same."* Etc. "Fourteenth." *"Can't be sure."* "Fifteenth." *"They are all crosses."* "Why?" *"If they weren't, some circles would have turned up."*

We see the differences in reactions between this stage and the preceding one. While the subjects of stage I spoke only in terms of empirical frequencies, these subjects immediately see the problem in terms of mixture and combinations: *"They cannot fall all with the same side up because while falling they turn over according to chance,"* Han said; and Ber, *"because they are too mixed."* Random mixture for these subjects has therefore, a combinatoric significance and is no longer simply a material mixing.

This is the reason for their very clear attitude when faced with the false counters. Because their earlier combinatoric intuitions allowed them to go beyond the initial level of phenomenism and egocentrism, they are no longer limited to explanations based on the power of the thrower, but consider the result of the throw immediately as contrary to the probabilities. They realize that the material itself is fixed, some immediately and others after a few hesitations. Han, after a few seconds, says, *"It's because there are crosses on both sides,"* which is what Ber said without even checking to see the other side of the counter. *"Perhaps there is a cross on the other side,"* guessed Bra, and he needed to examine only one piece to make a judgment about the whole group (contrary to what Wi of stage I had done). The subject Mal made the hypothesis after the second throw and was certain at the third one. In brief, whatever the modality of the discovery, it shows immediately a different orientation of mind from the first stage.

The lines of separation, however, between this stage II and the third stage are more difficult, and this is natural, since the difference which separates them is related to the progress of the operations of combination (see Chapter VII), and since the present experiment is not concerned with combinations as such, but only with their indirect results. What we can say in general, that is, from a comparison of different indices we have found (but without any precise clinical sign allowing us to make a decision on it), is that the child at this level does come to a global sense of the probabilities but does not achieve finer evaluations which suppose a progressive quantification. As for the question of knowing if the true counters could give the same results as the trick ones, the subjects answer correctly either that it could not happen, or that it could happen only rarely. Mar even goes so far as to predict that this would be more likely with ten thousand trials than with one thousand. For the question of knowing if the true counters will turn up more crosses or more circles or an equal number of each they can only answer that they do not know. They do not understand, therefore, that there are different degrees of probability, and that 50 percent of each is the most probable distribution. In addition, when it is a question of deciding from which of two sacks, true or fixed, we are drawing the counters, these same subjects try simply to guess, or are doubtful, and it is only around the tenth or fifteenth trial that they are suddenly certain, as Man. We will find more nuanced reactions during stage III.

3. *The third stage: Quantification of probability.*

In parallel with the combinatoric operations (Chapters VII through IX) the subjects of this last stage come to finer judgments on what are properly called questions of probabilities, some of the difficulties of which we have just seen in the preceding stage:

MUL (11; 9) claims that we cannot know the results of throwing twenty counters, whether there will be more crosses, more circles, or the same number of each because *"it's chance, it depends."* "And is it more likely to get half and half with one thousand or ten thousand throws?" *"With a million."* (False counters.) *"All crosses! You must have a trick someplace in here."* (He turns over a counter.) *"There!"* "But do they all have crosses on the other side?" *"No doubt about it."* "Would it be possible with true ones?" *"Yes, if we did it many times."* Guessing if the counters are being taken from the first sack or the sack of trick ones: "One." *"O.K., it's chance. Don't know."* "Two." *"If I don't know the first, I don't know the second."* "Three." *"If you continue one by one, you will not get a cross every time."*

TAM (12; 10): For one counter: *"It's chance deciding, but the more there are, the more chances there are of knowing if they are mixed, with half of each."* (False counters.) *"All crosses."* (He turns one over.) *"Oh, both sides the same."* "But could there be one true counter in the pile, one with a circle on the other side?" *"Yes, one."* "Two?" *"Yes."* "Five?" *"That would be unlikely."* Determination of which sack has true counters: "One." *"It's maybe a trick counter, but I'm not sure."* "Two." *"We can't be sure."* "Three." *"That's the third cross we've gotten. If we continue, we'll know. Two can happen, but three is unlikely."* "What can we do to be sure?" *"Continue. We're getting more and more sure."* "Why?" *"Because we see the number of circles and crosses, and if there are nothing but crosses, we'll know they are the false ones."* "Why are we more sure with five than with three?" *"Because with three, the chance is less great."* "And twenty compared with ten?" *"Same thing."* "With the true counters could we have only crosses come up?" *"No."* "However with five counters, we can get five crosses?" *"Yes, but the number is smaller* (than the twenty of sack 2)." "How many times would we have to toss them to get five crosses?" *"Twenty-five times."* "And with six counters?" *"Thirty-five to forty times."* "And with seven counters?" *"Impossible."*

SOM (13; 5): Twenty counters at once: "Will there be more circles than crosses?" *"It would just be chance to get more of one than the other."* "Could they all fall with crosses up?" *"No."* "If we make one thousand throws?" *"Maybe."* "Is there a better chance with more or with fewer trials?" *"With a large number."* Determination of which sack has the true counters: "One counter?" *"We can't know."* "Two?" *"Don't know."* "Three?" *"Still don't know."* "Four?" *"Probably the trick ones."* "Sure?" *"Not yet."* "And if we continue?" *"We will be much more certain."* "Why?" *"Because if there are a lot, it would be extraordinary that all come up crosses."*

In comparing these reactions with those of stage II, we see what the progress is: The intuition of probability has become susceptible to a continual growth of quantification on the part of the subjects. No doubt, as in all studies of the development of chance, there is no single and absolute index permitting us to establish with certainty the presence of such a characteristic. But the group of indices is significant in itself, and especially their convergence when we compare the results of different analyses (see in particular those of Chapter III).

In the problem of guessing from which sack we were drawing the counters one by one, the difference in reactions between this group and the preceding one is clear. In stage II the subject had the tendency to decide immediately (cf. Mal in Section 2). or if he does not decide it at the first counters, he doubts it up to the moment when he is suddenly convinced, without transition (see Man in Section 2.). We see, on the contrary, that the preceding subjects exhibit the continual growth of probabilities. Mul is content to sketch out this scheme, but Tam expresses its progression very well, *"If you continue, we will know . . . we will be more and more certain,"* and he says this because with small numbers *"chance is less great"* (the justification is even more significant). Thus it is for this reason that with five counters it could happen that once in twenty-five throws that the five crosses would show at the same time, but with six counters, we would need thirty-five to forty throws, and with seven, it would be *"impossible."* However arbitrary these approximations might be, they give a remarkable sense of the increase of probability which leads toward that state of certainty when, according to Som *"we will always be more sure."* In comparing these reactions with the reflections of subjects at the corresponding stage in Chapter III (Section 6) concerning the law of large numbers, we notice the same mechanism at work in the present responses.

As for the question of knowing how the twenty or more true counters will come out when thrown all together, we note a similar progression over stage II. The subjects of stage III, in fact, come again to a sort of continuity in the increase of probability. The most probable situation is half crosses and half circles, but it is all the more probable as the number of throws increases (cf. the million of Mul). Inversely the least probable situation is to get all crosses or all circles, but even this situation becomes possible with a large number of trials (cf. Mul also).

In short, instead of probability remaining global as in stage II, or all of a piece, it is broken into gradual judgments which reflect the existence of an implicit quantification. We will try to examine this quantification in the course of Chapter VI, but with new methods. On the other hand, the continuity of this graduation marks an increasing relativity of probability as such. This relativity (in the sense of a combination of different complex data at play in opposition to simple attributive judgments) certainly constitutes the end result of the development of diverse combinatoric operations whose role we have already studied in the course of the preceding chapters, and which we will study in Chapters VII to IX.

4. *The experiments with marbles.*

The make-up of the experiment with marbles—whose results we will describe rapidly as a counterexample—is just about the same as the make-up of the preceding one, with a single difference, however, which will allow an easier understanding of the delivery of marbles of a single color. First we present the child with sack A containing twenty blue marbles with twenty red ones mixed in and ask him several questions about how they will come out. Then we give him sack B with only blue ones. In the case of the false counters the difficulty was to understand that the counters which were similar in appearance to the first ones had in reality crosses on both sides. In the case of the sack of blue marbles, there is, on the contrary, simply the absence of one of the expected elements. This makes it easier for the subjects to conclude either that there is insufficient mixture or that only one color is present. The correct answers appear, therefore, sooner because of a difference which is easily explained, thanks to the greater intuitive transparency of the data. But the successive order of the three types of responses is the same, and that is what we want to establish.[1] On the other hand, with those who discover the trick more quickly than with the counters, it is interesting to determine the context as well as the reactions to an additional question concerning the numbers present. Here are reactions of stage I:

GUE (4; 4): Mixed marbles. We show him the colors and we shake the bag: "If you take one of them what color will it be?" *"Blue."* (True.) "And another one?" *"Blue."* (Again true.) "And

[1] To keep it short, we will limit ourselves to an examination of stages I and II.

another one?" *"Red."* "Can you be sure?" *"Yes."* (Again he takes out a blue one.) "And what if we take out a handful?" *"Blue and red."* (Experiment: true.) Sack B (only blues). We shake it: "What if you took out a handful?" *"They will be red and blue."* (He takes out a handful.) *"All blues."* "Why? . . . What if we do it again?" *"They will be red."* "Nothing but reds, or red and blue ones?" *"Nothing but reds."* (Experiment.) *"All blues."* "Try again." (Successive experiments.) *"All blues."* (He seems to find this quite natural.) "Why are they always blue? . . ." (We take sack A again, show him the mixed contents, and shake it.) "If we take out a handful, what colors will we get?" *"All blues."*

EIS (4; 7): Sack A: "If you take out a handful, what colors will you get?" *"All blue ones."* "And then if you take another handful, what will they be?" *"All red."* "And a third handful?" *"Both colors."* "Try it." *"Both."* "Try again." *"Both."* Sack B: "What will come out?" *"Both."* (Experiment.) *"Nothing but blues."* "And now?" *"Nothing but reds."* (He tries twice in a row.) *"The red ones don't want to come out."* "Why?" *"The blues are pushing the red ones aside."* "Try again." After several tries, he says, *"There are only blue ones."*

RAY (5; 6) predicts blue and red alternately for sack A but is wrong every time. "Why were you wrong?" *"A person can be wrong."* "Now?" *"Red."* "Sure?" *"Yes, because this time I can't be wrong."* (Experiment: blue.) "Why?" *"I wasn't careful enough."* Sack B: "What will come out?" *"Both."* "Try." *"Nothing but blues."* "Try again." *"Now I know what it is. There are only blues!"* We put fifteen red marbles and ten blue ones in sack A, and have him count them:: "How many?" *"Fifteen red ones and ten blue ones."* "How many of each color will you get if you take a handful?" *"Same number of each."* "Why?" *"Because that way I won't be fooled, not this time."* (He takes out four reds and three blues. . . .) *"I was almost right."* "Now?" *"Same number of each."* (He takes out four reds and two blues.) *"Two more of the red ones, but almost the same."* "Why do you always get more of the red ones?" *"I don't know."* (We empty the bag and have him count them again.) "If you take a handful, there will be more of which color?" *"Same number of each, or probably more red ones, or even I think there will be more blue ones."*

Fin (6; 1): Sack B, after the third handful: *"There are only blues."* Sack A with fifteen reds and ten blues, counted by the subject: "We have more of which color in the bag?" *"More red ones."* "If you take a handful, you'll have more of which ones?" *"I don't know."* "And can I know?" *"No."*

Rob (6; 6): Sack A: "Can we know what you will get if you take out a handful? Will they be all reds or all blues, or mixed (we shake it well after showing him the contents)?" *"We can't know."* "What do you think?" *"All reds."* "Watch." (He gets more blues than reds.) *"More blues."* "Why?" *"Maybe it's because there're more of them."* "What was in the sack?" *"Same number of both."* "If you take another handful, what will you get?" *"More blues."* (He gets five red ones and six blues.) "And next?" *"More blues."* Sack B: he takes a handful: *"All blues."* "Another handful?" *"Again eleven blue ones!"* (He is surprised.) "Why?" *"Maybe the sack is full of them."* "Sure about it?" *"No."* "Another handful?" *"All blues."* "Could there be a red one in the sack?" *"Yes."* "Ten red ones?" *"Yes."* "And could all those still in the sack be red?" *"Yes."* Sack A: Ten blue marbles and fifteen red ones, counted by the subject: "If you take a handful, will you get more blues, or more reds, or the same of each?" *"More blue ones."* "Why? . . . Can we know?" *"Yes, if we think hard about it."* "Next?" *"More blues."* "What's left in the sack?" *"More red ones."* "Then if you take a handful now, what will you get?" *"More red ones."* "Will you be more likely to get more red ones if you take a large or a small handful?" *"A small handful is better."* (He takes out a few.) *"No, a big one."*

We note that the youngest subjects of this stage, that is from four years to about five or six, are fooled by sack B as were the subjects of Section 1 with the counters with crosses on both sides. They do not conclude that the sack is filled with only blue marbles, or they do not explain how they are getting only blue marbles, or they give egocentric reasons (*"the red ones don't want to come out,"* or *"the blue ones are pushing the red ones,"* etc.). The older subjects, on the contrary, understand very quickly that the sack is filled solely with blue marbles, and the reason for this, we repeat, is that it is easier to grasp intuitively the absence of red marbles than to imagine a counter with crosses on both sides. But what is the meaning of this comprehension, and does it constitute a reaction of stage II, or is it simply a substage I B, because along with it go some typical reactions of stage I?

There are several interesting points we can note in this regard. In the first place, the child frequently imagines, by means of a very revealing contradiction, that "we can't know" which color marble we will get from sack A, but that nonetheless *"we can, if we think hard,"* as Rob said. Now, thinking hard in no way means being able to conceive of the possible combinations due to the mixture and understood as a function of the given frequencies. The child at this level does not believe that the physical mixture in front of him excludes the possibility of actions or hidden influences that the elements may have on each other. Thus he simply tries to reconstitute the elements, admitting that each drawing acts on the next one, either in the direction of compensation (which is natural and has been discussed already in Section 1) or in the direction of repetition (which is quite curious). Thus Rob, having observed more blues at one drawing, predicts several times, with no hesitation, *"more blues,"* even when we were now playing with fifteen reds and ten blues in a new game. What is especially striking is this indifference of the child to quantification. In spite of his having counted the fifteen red marbles and the ten blue ones which we put in sack A at a given time, he is not concerned with the numbers when it comes to predicting the composition of a handful. *"The same number of each,"* Ray, for example, said, *"because I am not going to be wrong this time."* And since the experiment proves him wrong each time, he ends up saying that he will get *"more blues"* after he remembered that there were more red ones in the sack.

In short, it is not an exaggeration to conclude that these children understand nothing about the notion of random mixture. To come back to the question of sack B (with blue marbles exclusively), we see that their reactions to it do not reach the level of stage II. First, we note that even before this experiment, Eis and Rob, for example, take as probable that a handful of marbles from sack A (mixed) will be *"all blue"* or *"all red."* On the other hand, when they can get only blue ones from sack B, several subjects, even after checking the energetic shaking of the sack, decide nonetheless that the next try will give *"all red ones,"* as if the two colors of mixed marbles had been separated in the sack as the hand seized a bunch. This is the way that Eis and Ray react; and Rob, after taking some handfuls of marbles which were all blue, admits that the rest are maybe all red ones. And he says this even after he guessed that the sack could be filled with just blue marbles.

In these conditions, it is already clear that for the majority of the subjects of level I B the discovery of the exclusively blue composition of sack B does not constitute an unequivocal index of the notion of combinatoric mixture and consequently the notion of chance. In reality it is only a question of an elementary inductive process related to that induction which we called (Section 1) passive, or empirical, to distinguish it from active or experimental induction. The child simply concludes blue follows blue as he admits that one drawing will let him see something about the next. Still this fact is not unworthy of notice: On the one hand, it confirms the precocity of intuitions of frequency and rarity, while showing, on the other hand, that they are an insufficient basis on which to construct the notion of chance, because this notion supposes the notion of a random mixture which is not only empirical (material mixture) but combinatoric.

It is this fact that the subjects of stage II will prove for us now. Following are three examples, starting with an intermediate stage between levels I B and II:

GER (6; 9): Sack A: "What will the first handful look like?" *"Red ones and blue ones."* "And the second?" *"Same."* Etc. Sack B: "Take a single marble. What color will it be?" *"Red."* (Experiment: Three tries in a row give three blue marbles.) *"That's because there aren't as many red ones as blue ones."* "Take a handful." *"All blues; that's because ther're many blue ones and few reds."* "Again." *"All blues."* "What are the red ones doing?" *"Maybe they're all at the bottom."* "Another handful." (He gets four blues.) *"There are none!"*

WEI (7; 4): Sack B: "Take a handful." *"All blues."* "Why?" *"Because."* "When you take a second handful, what color marbles will you get?" *"Blue ones."* "No red ones?" *"No."* "Why?" *"Because there aren't any."* "Try." *"What a lot of blues!"* Sack A: Fifteen red marbles and ten blues: "And if you take a handful?" *"More red ones."* "Why?" *"Because there are more red ones in the sack.*" (Experiment: four reds and six blues.) "*How's that! More of the blues.*" "Why?" "*Because I didn't take a lot.*" "What if you had taken more?" "*There would be eight red ones and six blues.*"

FRAN (7; 6): Sack B: "Take a handful." "*All blues.*" "The next handful, what will it be?" "*Blues and reds.*" "Why?" "*If they are well mixed.*" (Experiment.) "*No, there were only blue ones in that sack.*" Sack A: Fifteen red ones and ten blues, which the subject counted: "*We'll get some reds and some blues, but more reds.*" "Why?" "*Because we took away some blues.*"

We see the difference between these reactions and the reactions we saw in stage I. The child is not content with simply understanding that sack B contains only blue marbles, he already calculates the probabilities as a function of the quantities (especially with sack A holding fifteen red ones and ten blues). Thus, at first Ger thinks, as he takes the first blue marbles from sack A, that "*there are many blue ones and few reds,*" then he calls it a bad mixture, and finally he concludes that all of them are blue. Wei, especially, gives a remarkable explanation for his error in predicting that he will take more red ones than blue ones from the sack of fifteen reds and ten blues: he says that his sample is too small, and that a larger handful will give him a better chance of proving his prediction. In short, there are two new elements in these answers which set them apart from the preceding ones: the idea of combinatoric mixture (Ger and Fran) and a beginning of of the quantification of probability.

Appendix.

Related to the intuitions of frequency and rarity which we have seen occur in the course of the preceding answers, we also took the time to question the subjects used in this chapter about what they understood by the word *chance* or the expression *by chance.*[1] It is clear that a verbal notion does not tell us about the actual

[1] See also Marguerita Loosli-Usteri, "La notion du hasard chez l'enfant," *Archives de Psychologie* 23 (1931), no. 89.

operations of thought, but it does constitute a conscious grasp, more or less adequate, of the results of the operations. It is in this way that we conceive it. In addition, and precisely because it is a conscious grasp, it always occurs later than the operations themselves, but its evolution often reproduces the evolution of the operations, even though partial and imperfect.

In the case of the verbal notion of chance, we find actually three sorts of interesting reactions: first, the reactions of the subjects who do not understand the word *chance* (up to about six or seven years); second, the reactions of those subjects who give it the sense of a rare or exceptional event (six to nine years); and, finally, the correct reactions of those who define chance as the interaction or interference of independent causal series. As for the technique adopted, we sometimes told a little story asking the subject if it was a question of chance or not, and especially we asked the child to tell a story or event which depended on chance.

Here is an example of failure to understand: "What does 'chance' mean?" *"It's when something happens real quick. I can't say it another way."* "For example?" *"If there would a be a fire once."* (MIC, 6; 3.)

Here is an example of definition by rarity: *"By chance I did something wrong."* "Why by chance?" *"Because it does not happen often. Ah, if that happens often, then it is not chance."* "Give me another example." *"By chance, I am sick."* "And if I say that 'an auto knocked me down by chance,' is that correct?" *"Yes, because that doesn't happen every day."* (BON, 9; 1.)

Here finally are two examples of definition by interaction or interference: *"Chance: if he has been invited for the first time, and he falls ill on just that day."* (BOB, 9; 11.) And: *"I didn't know whether I wanted to go see my grandmother or my aunt, and so I went to see my grandmother and found there my aunt."* (PAM 13; 1.)

Without insisting on the significance of these verbal definitions, we have thought it useful to point them out in the light of the previous reactions.

V. The Random Drawing of Pairs

The results which precede (Section 4 of Chapter IV) have made us realize the importance of the problem of the progressive quantification of probabilities in the development of notions of chance. It is only proper, therefore, to take up the question systematically, which we will do in the following chapter. We would first like to describe an experiment which already has a bearing on quantification, but which will allow us to tie together the earlier analyses which up until now have led us to the combinatoric concept of random mixture.

We put out on a table a large collection of elements A (for example, fifteen yellow counters), a smaller set B (ten red counters, for example), a still smaller set C (seven green counters), and a very small set D (three blue counters). We leave these sets of elements on the table so that the child will have no difficulties remembering the colors and proportions, and we put duplicates of these four sets in a sack and mix them well. The child then is instructed to put his hand in the sack and bring out a certain number of pairs of the counters. We ask him to predict the most probable pair before each drawing. (As pairs of the counters are drawn they are naturally put on the table in front of him so that he can know what still remains in the sack, but we give him no explanations about this.) With the little children, we can also have them predict the drawing of a single counter at a time.

The results obtained match exactly what we had seen earlier. In the course of stage I, the child seeks to make predictions as a function of the possible number of combinations, but always with the same sorts of criteria, one of them being the number of elements in play. In the course of stage II, there is a search for quantitative relationships, but without understanding the need to remake the calculations after each drawing as the number of elements remaining in the sack diminish precisely by the number already drawn out. In the course of stage III, finally, the probability is quantified as a function of the counters remaining in the sack.

1. *The first stage: Absence of systematic probability.*

First here are some examples concerning the drawing of isolated elements:

BER (5; 1): Six reds, two blues: "If you take a counter out of the sack, but without looking inside it, what color do you think it will probably be?" *"Red."* "Why?" *"Because I like red a lot."* (Same question, but this time using six blue ones and two red (ones.) *"Red."* "But there are lots of blue ones. Don't you think it will more likely be a blue one?" *"Yes."* (Six blues and one white.) [1] *"White."* "Why?" *"Because white is the first."* (He points to the scale of examples on the table.) "But we mixed it all up in the sack as you saw." (We show him by mixing the pile on the table.) "Now the white is no longer the first. Take one of them from the sack. What color do you think it will be?" *"White."* "Why?" *"Because you mixed it."* (Six white ones and one red with corresponding sets on the table, as we always do.) *"Red."* "Why?" *"I like red."* "Look at what you got." *"White."* "Why?" *"Because it was in the corner of the sack."*

LEU (5; 8): Six reds, four blues, one white: The counterparts

[1] We naturally repeat each time the entire series of questions, varying the words.

are lying on the table and Leu himself shakes up the sack. "If you take a counter out of the sack, what color will it be?" *"White."* "Why?" *"Because there is only one white one."* (We take the white one out of the sack and also the one from the table.) "Now if you take one, what color will it probably be?" *"Blue."* "Why?" (He opens his hand.) *"Red!"* "Why is it a red one and not a blue? Is there a reason?" *"No."* (Six reds and one white.) *"Red."* "Why? . . ." (Four white ones and four reds.) "What will you get?" *"Red."* "Why? Is it more likely that you'll get a red one?" *"Yes."* "Why?"

MON (5; 10): Six blues and three reds: *"Red."* "Why?" *"Because."* (Six blues, two reds.) *"Blue."* "Why?" *"Because."* (Three blues and six reds.) *"Blue."* "Why?" *"Because."* (Three blues, six reds.) *"Blue."* "Why? . . ." (Twelve reds and one white.) *"White."* "Why?" *"Because there's one white one."* (Twelve white ones and one red.) *"Red."* "Why?" *"Because you put one in."*

SCHA (5; 10): One rose, two yellow, and five red ones: *"Red."* "Why?" *"Because there are a lot of them."* (We do it again adding six blue ones.) "Now?" *"A yellow one."* "Why is a yellow one more likely?" *"I don't know."* (One rose, two yellow, two red, and six blue.) *"I think that it is the rose that will come out."* "Why? . . ." (Four blues and four yellows.) "Can we know what will come out or not?" *"Probably a yellow."* "Why?" *"It's prettier."* (One blue, two yellows, three greens, four reds, and five rose.) "If you just take any two of them, what color do you think they will be?" *"Yellow."* "Why?" *"Because there are two of them."* (We use four yellows, four reds, and four rose.) "And now?" *"Red and rose."* "Why?" *"Because that's pretty."*

LAU (6; 6): Two blues and six reds: *"Red."* "Why?" *"Because there are more of them."* (Two white ones, five reds, and five blues.) *"White."* "Why?" *"Because there aren't many of them and there are the same number of reds and blues."* (One blue, three reds, and six whites.) *"White."* "Why?" *"Because there are more."* (Three rose ones, three reds, and two white ones.) *"Rose."* "Why?" *"Because there are three rose ones and three reds."* (Same plus four green ones.) "Now?" *"Red because there aren't many of them."*

FIN (6; 4): Two whites and eight blues: *"A blue one because there are more of them.*" (Four whites and four reds.) "*We can't know because there are as many white ones as red ones.*" (One blue, two greens, two rose, and three reds.) "*A red one because there are more of them.*" "And next what is more likely, a green or a rose?" "*A rose because that's right next to red* (i.e., as a color or on the table given as a memory help)." (One blue, two green, three rose, four red, five blue, and six yellow.) "If you take two, what colors will they most likely be?" "*Two yellows because there are more of them.*" (And there are fifteen others!) "And if it isn't yellow?" "*Two blue ones.*" "And when those are taken away?" "*Two reds.*" "If we choose at random, can we get different colors?" "*Then a rose and a red* (i.e., as neighboring colors)."

PER (7; 3): One white, two rose, three green, four red, five blue, and six yellow: "If you choose at random, any old way, what are you likely to get?" "*White.*" "Why?" "*Because that's first.*" "But if I mix them up well, it is still first?" "*Yes.*" (We take them all away and mix them up again.) "*A rose one.*" (Six blues and three whites.) "*A blue one.*" "Why?" "*Because there are many of them.*" (Six whites, five yellows, four blues, three greens, two reds, and one rose.) "If you take two of them at a time, what will you most likely get?" "*Red ones.*" "Why?" "*Because there are two.*" "What if we mix it up well?" "*White and red.*" "And then without those?" "*Red and yellow.*" "And then if it is well mixed?" "*Then white and rose* (i.e., the two extremes on the memory help)."

These are the usual reactions for this first stage. There is one general trait which, above all others, is striking and which confirms with singular clarity what we have constantly noted at this level during the previous experiments: Everything happens as if the young subjects did not take into account the mixture of the counters in the sack, in other words, did not understand the nature of random mixture as a system of fortuitous interactions. It is for this reason that we put on the table in front of them the same sorts of pieces that we put into the sack, to act as a memory jog, and arranged on a scale according to the color (i.e., one white, two green, three rose, etc.). In this way the child could see the differences without having to count them. Now, we found that many of the subjects, even though they knew that the pieces in the sack were mixed (not only did we repeat this endlessly, but the children themselves shook the sack to mix the contents), made their decision as if the order of the model scale was going to control the results of their drawings, in other words, as if the pieces, even though mixed up in the sack, were going to obey the order determined by the initial model. For example, Ber started out by predicting that, out of six blue pieces and one white, the latter would be the likely one to be drawn when drawing a single one. The reason he gave was that it is *"the first* (in the model)," and he took no account of the numbers present. When we insist on the mixture by indicating that white is no longer the first one, Ber again calls for white, but this time *"because you mixed them,"* as if its initial order had not been changed but, on the contrary, allowed us to get it from among the mixed elements. The first criterion which the child follows in making his choice is, therefore, the initial ordering of the elements.

A second criterion, equally frequent, again shows the same failure to understand the idea of mixture and the complete neglect of the influence of quantity. When we tell the subject that he is going to draw out a single piece, or any two pieces, it often happens that he predicts that the color he will get is exactly the one which is shown the same number of times on the model scale—that is, the single piece, or the color shown by two pieces. In other words, his decision is not at all based on the fact that the color which has the greatest number of pieces has the best chance of being drawn; on the contrary, he predicts what color will be drawn by seeing which color on the model has the same number of pieces he is asked to draw. It is in this way that Leu, out of a collection of six reds, four blues, and one white, predicts that he is going to get the white one *"because there is only one white one."* Mon, with one white and twelve red ones, predicts that he will get the white *"because there is one white one";* and out of a dozen whites and one red, predicts red *"because we put one red one in."* Scha, in the same way, asked to take a pair at random, says two yellow ones (out of fifteen elements) *"because there are only two of them* (on the model)." In the same way, Per with twenty-one pieces believes that he is going to get a pair of reds *"because there are two of them."* It is, no doubt, this argument that controls some of the strange answers occasionally gotten which give predictions based on the small number of pieces shown. Thus Lau, after giving a seemingly correct answer for a group of two blues and six reds, *"Red because there are more of them,"* makes a choice given two whites, five reds, and five blues, *"White because there aren't many of them, and there are the same number of reds and blues."* When we want just a single piece, the child seems to say, it cannot come from the more numerous groups, especially if they are of equal number, since there is nothing to determine his individual choice; while with two white ones, he has no choice but to take one of the two.

A third type of criterion is closely related to intrinsic qualities of the elements about which predictions are to be made, and this is always done without consideration for the effects of mixture or the quantities at work. It is for this reason that Fin, who mentions, however, at the beginning the number of elements, thinks that if two pieces are to be drawn, not of the same color, that they will be *"red and rose"* because these are two colors which are close together. Scha in the same way calls *"red and rose because that is pretty,"* that is, that these colors go well together.

A fourth criterion is the subjective preference, as if chance were equivalent to simple arbitrariness. Ber predicts *"red because I like red a lot,"* and Scha says, *"probably yellow, it's prettier."*

The fifth criterion, finally, shows a more precise idea of probability: This is quantity itself. Of two sets, the more numerous one has a better chance of coming out in the drawing, the child seems to say. But we must be wary of an interpretation so close to our own view of things, precisely because of the fact that this last criterion is constantly mixed up with the other ones. To be precise, we must distinguish two cases of quite different meanings, with all the intermediary steps between the two extremes. In the first case —the one in which we find the youngest subjects—in the eyes of the child, quantity is simply one quality among several: In the same way that one piece may be chosen because it is first, or because it is shown by a single example on the model, or because it is pretty, and so forth, it also may be chosen because there are many of that color, without, however, there being any understanding of the effects of mixing. Scha, for instance, with one rose, two yellows, and five reds predicts *"red because we have a lot of them* (note that he says 'we' and not 'there are')," but when we add six blues, he predicts yellow for no reasons and next moves on to subjective criteria. The second case, on the contrary, consists of subjects who seem to be moving toward a correct concept of quantification seen as one of the aspects of intuition of the effects of mixture, but who, at the same time, fall back on other criteria as they see the need. Fin, for example, begins with answers which are close to what we get at stage II: *"Blue because there are many of them,"* and so on; but with one blue, two greens, two roses, and three reds, he predicts *"a rose one because that's next to red,"* and this is simply a return to the criteria of rank, or selected similarities.

We note now that this second case becomes more common near the end of this stage, showing quite clearly that it is a question of a growing intuition of probability which will develop in the course of stage II. In this case we can, therefore, distinguish, as we have done occasionally before, a substage I B as an intermediary between stages I A and II.

This description leads us to two conclusions. The first one, as we have been able to note already several times, is that the idea of mixture, which is the starting point for the idea of chance, is far from being as primitive and as simple to grasp as it would seem. When we say mixture, we are saying, in fact, more or less complete independence of elements which come into contact because of interacting trajectories (this is certainly why mixture is the prototype of that interaction between independent series by means of which Cournot defined chance). Subjects of stage I reason, on the contrary, as if once the elements were mixed, they kept the characteristics which they had prior to their mixing—i.e., they kept precisely what the mixing destroyed. For example, the marbles which are shaken up in the sack are expected to keep the same order that they had originally, the order which is shown on the table used as a memory guide for the child. These strange reactions recall an observation we made some time ago with A. Szeminska: For certain young subjects, a set A appears larger than a set B when set A is taken from a sample larger than the sample from which group B was taken (for example six pieces taken from a pile of thirty pieces will be more than six other pieces taken from a pile of ten). In other words, mixture, as it is conceived at this level (under seven to eight years) is not yet irreversible. This fact is all the more paradoxical because the child at this level is neither capable of rational observation nor of operative reversibility—which we already noted concerning stage I in Chapter I (in the course of which the subjects attributed to marbles mixed by a tipping movement the ability to come back to their original places). Before the notions of irreversible mixture and reversible operations are developed together in stage II, what we have is only a world of perceptive and subjective intuitions (phenomenism and egocentrism) whose qualities are conserved by simple transfers or by preservations. This is the contrary of that intuitive pseudo-conservation which was only arbitrary and not yet a true understanding of chance.

Now, if the notion of chance derives from the intuition of random mixture, the judgment of probability will consist of anticipating the possibilities of recovering the mixed elements, such an anticipation being based solely on the quantitative relationships at play. The second conclusion to be drawn from the preceding facts is then, that lacking such a notion of irreversible mixture, the child at this level will not come to the idea of structuring his probabilistic evaluations by grounding them on a systematic quantification. On the contrary, he will not even have the power to do this since such quantification would suppose a system of operations which were at the same time numerical and combinatoric, precisely the absence of which explains why the development of the concept of random mixture remains unrealized.

When Mon, for example, with twelve white counters and one red one shaken up in a sack, immediately falls into the decision of choosing the red one because it is single, he shows not only a weak sense of mixture but also a very rudimentary notion of probability and of the operations necessary for finding certain elements in a mixed set. In this particular case, in fact, when a subject reaches for a counter in a mixed sack, the judgment of probability consists in anticipating the probability of his hand finding the desired elements. We see then how this anticipation leads necessarily to operations of quantification. To be certain of getting the red counter, all that is necessary is to get the whole set of counters in his hand, that is, to pull out the thirteen elements of the thirteen present. This would give a certainty, or a probability of 13/13 which equals one. To get a particular one of the elements, it is a question of recognizing it as a fraction of the whole, that is, as one part to be compared quantitatively with the rest of the set, which amounts to expressing certainty (value of one) in fractions corresponding to the pieces desired. In expecting that it was probable that he would get the red counter, Mon reasoned as if his hand were not exposed to twelve times as many contacts with white counters as with red ones. That is, it was not as if the single red counter was simply the thirteenth element of a whole considered as a set of elements. He, therefore, attributes a sort of direct correspondence between his hand and the red counter considered absolutely and as a unit, instead of seeing it fractionally as two probabilities: 1/13 or 12/13, which would give the greatest probability to the white counters. In other words, he has as yet no notion at all of probability which the subjects at level I B are already oriented toward, since they do talk about the number of elements present as a characteristic of the particular element desired. They are becoming oriented toward true probability precisely because they divide the total group into categories, recognizing quantified relationships between the pieces, instead of reasoning about a particular element or isolated quality.

Psychologically as well as logically, a judgment of probability goes hand in hand with the development of quantitative operations of combinations. While chance is the result of an irreversible, random mixture, the judgment of probability consists, on the contrary, in a reduction, by means of thought processes, of the mixture to a system of combinations that can be expressed as a fraction.

2. *The second stage: Beginnings of quantified probability.*

This quantification of probability begins at stage II as soon as the proper idea of random mixture is developed. But the subjects of stage II, instead of taking into account after each drawing the number of elements in each category which remain in the mixture, forget to count how many counters they have already taken out of the sack. Therefore, they reason on fixed categories as follows:

FRE (8; 6): Eight yellow, four red, two green, and one blue: "If you take out two at a time, without looking, what colors will you most likely get?" *"Two yellow ones."* "Why?" *"Because there are a lot of them."* (He gets two of them.) "And if you take just one now, what color?" *"Yellow again."* (He gets a red one.) "And now?" *"A red."* (He gets a blue one.) "And now?" *"A green one."* "And then?" *"A yellow."* "And next?" *"A red."* "And next?" *"A green."* He follows thus the order of colors according to a decreasing rank of frequency.

PIE (9; 3): Eight rose, six yellow, four white, and one green: "If you take two, etc." *"Rose and green* (i.e., the extremes)." "Try it." (He gets a rose and a white.) "Why a rose one and a white?" *"There are a lot of them."* "And now?" *"White and yellow."* "And next?" *"Green and white."* (He has gone right down the scale.)

DAN (9; 6): Eight red, five yellow, two rose, and one green: He calls for, in succession, pairs of red/yellow, red/rose, red/green, linking in this way an element of the largest group with an element of each of the other groups in a decreasing order of frequency.

GIN (10; 0): Fifteen green, twelve yellow, eight red, and three blue: "If you take two out, what colors will you be most likely to get?" *"Two greens or a green and a yellow."* (He gets two greens.) "And now?" *"Green and yellow."* (He gets yellow and blue.) "And now?" *"Two reds."* "Look at what you've already taken out: Is there a better chance for green or for red?" *"Red because we've already gotten two green."* (But there are still thirteen left!) "Try it." (He gets two green ones.) "Why did you get two green ones again?" *"Because there were more green ones than any other."* "And now?" *"Green and red."* (He gets green and blue.) "And now?" *"Two yellows."* "Why?" *"Because we haven't gotten many."* "And now a blue?" *"I don't think so."*

CHEV (10; 0): Eight yellow, six red, four blue, and two green:

"Yellow and red." (He gets two red.) "And next?" *"Two yellow."* (He gets a yellow and a red.) "And now?" *"Two greens."* "Why?" *"Because we haven't gotten any yet."*

The meaning of these reactions is clear. On the one hand, all these subjects take quantity into account, showing quite well that they do understand the effects of mixture and also that a hand put into the sack has more of a chance of getting a piece of one type if that type is the most numerous. However, if these notions are correct at the beginning of each experiment they are not so valid as the experiment proceeds. Instead of taking into account what remains in the sack after each drawing and adjusting the quantified relationships accordingly, the subject continues to follow the order of decreasing frequencies based on the initial numbers or even forgets the inequalities and reasons by categories as though they were equivalent.

It is in this way that certain subjects, even though starting from a known quantitative distribution, still move from one extreme to another, like Pie who predicts as the most probable pair rose (eight pieces) and a green (one piece), and Chev who calls for two greens at the third drawing (two out of twenty) because *"we haven't yet gotten any of them."* This first method recalls the processes of the first stage, although allied with a clear concern for quantity.

A second method consists in following the decreasing order of frequencies, like Fre who establishes a kind of rotation of turns, or Dan who associates for each pair an element of the most numerous group with an element of each of the others in a decreasing progression. It goes without saying that this second method, even though based on a principle of quantity, does not take into account the exact relationships between the frequencies.

The third method, finally, consists in starting with the most numerous groups, but next predicting simply on the basis of the colors not yet drawn instead of taking into account the numbers of the colored counters remaining in the sack. Gin, for example, predicts for the third pair *"two reds because we already have two greens* (the remaining mixture contains thirteen greens and eight reds)," and later predicts two yellows *"because we haven't gotten many of them.*" This third procedure, even though neglecting the relativity of the changing mixture, leads nonetheless unconsciously to a situation taking into account the remaining quantities.

It is striking to note the analogy between the judgments of probability and the methods of combination which we will study with the same subjects in Chapter VII. In the two cases the child is looking for an object system instead of proceeding simply empirically and on isolated constructions. But in both cases the system falls short because the subject conceives the whole of the probabilities as well as that of the combinations as a static sequence instead of understanding the dynamics of the interactions and successive modifications.

3. *The third stage and conclusions.*

The subjects of stage III succeed in quantifying the possibilities again after each drawing, showing in this way that they are able to conceive the relationships at play as being modified after each transfer of elements:

STO (10: 0): Twelve whites, ten reds, five greens, and three blues: *"A white and a red."* (He gets two greens.) "And now?" *"White and red."* (He actually gets these colors and continues in this way for several trials.) *"We now have less chance than before of getting white and red because we've taken out a number of them. More likely it will be a green and a blue."* (Twelve roses and six reds.) *"There is more chance for two rose ones because there are twice as many of them."* (He gets two rose.) *"There is still a better chance for the rose but less than before."* (He gets them.) *"Now there is almost the same chance for both."*

LAUR (12; 3): Fourteen greens, ten reds, seven yellows, one blue: *"Two greens."* (He gets a red and a yellow.) "And now what will it be?" *"Two greens."* (He gets them.) "And now?" *"There is still a better chance for two greens, since twelve of them remain, and there are only nine reds and six yellow."* "Try it." (He gets green and blue.) *"Hey! there's the single blue one."* "And now?" *"Still the two green ones are the best chance."*

As simple as these answers are, we see that the real difficulty for the subject consists in making an adjustment after each trial for the number of pieces still in play (in the same way with the systematic method of combinations, the difficulty will be to coordinate the combinations of each color successively with all the others). It is a case then of constantly coming back to the current situation instead of considering the terms of the problem as static or independent of each other. What is happening here is comparable to what is called a modification of systems of reference during a motion to which they refer.[1] This is why the intervention of formal operations seems necessary in this case.

The essential conclusion to be drawn from the preceding observations is that the notions of chance and probability are by nature essentially combinatoric. This is why we studied the development of concrete operations of combinations with the same group of subjects. But without yet knowing the details of this operative construction (which we will describe in the course of Chapter VII), it is interesting to try to discern from the present facts the increasing intervention of combinatoric schemes in the spontaneous reactions of the subjects.

It is thus clear that in the course of stage I such schemes are not shown yet at all; there is, however, nothing surprising in this since, before about seven years of age, the child is not capable of operative seriations or nesting of classes, nor certainly of an understanding of elementary numerical operations. It is nonetheless clear that from level I the subject possesses certain empirical intuitions which give him the ability to anticipate what will later be knowledge of chance. These empirical observations could already allow for a combinatoric interpretation, but the child does not get to that, and it is precisely because he lacks such an operative ability to structure that he fails to construct the notion of chance. Thus he does know of the existence of random mixture as a material mixing. Sometimes putting his playthings together and other times piling them up pell-mell, he knows quite well that finding a particular object is easier in the first case than in the second. And in this second case, when he wants to find what he has lost, he

[1] See, for example, Jean Piaget, *The Child's Conception of Movement and Speed* (New York: Basic Books, 1969).

sees quite well that his success depends on the number of elements and the different possible associations between them. He constantly has the chance to experience directly the same kinds of fortune—good or bad. However, and here is what the facts studied in this chapter teach us, the intuition of a mixing of concrete objects is not yet the understanding of a combinatoric mixture; and without an active ability to structure experience, the intuition does not suffice, by itself, to lead to the idea of chance, nor certainly to the idea of probability.

In fact, what do we get from the pure experiment of presenting a mixture, or from what is its equivalent, from a system of positions or unpredictable displacements (or, in even a more general way, from a system of unknown causes and multiple effects)? This simply yields observations of frequency: Certain associations are often produced while others are, to varying degrees, less frequent. But if this intuition of frequency constitutes a beginning of the intuition of random mixture, it is only an incomplete and gross approximation. It is quite compatible with other elements which are foreign to the concept of random mixture and which contradict it. And this intuition of frequency even calls for these other elements when adequate operative schemes are lacking. We have seen how the subjects of stage I, in spite of mixture, believe in the intervention of qualities or intrinsic powers in mixed objects (rose attracts red, etc.) and even in relationships that are inherent in the order which these objects had before being mixed up (the order, for example, of the colors on the desk used as a memory aid). To know that the subject has succeeded in getting an exact notion of random mixture, it is, therefore, necessary that two conditions be fulfilled, both of the conditions going beyond the pure experiment and the intuition of frequencies: We must have a table of the possible combinations, and only these give significance to combinations actually observed; and we must also have a system of quantitative relationships among the different categories which are being combined with each other as well as the list of all the possible combinations, that is, the sample space. In both cases, therefore, it is necessary that a system of operations be developed which goes beyond the data and which can assimilate this system with properly operative schemes.

We have seen how the passage from stage I to stage II, that is, from the simple intuition of empirical frequencies subjectively interpreted to the idea of random mixture and the beginning of correct probabilistic estimates, was precisely characterized by the development of certain combinatoric schemes. In the first place, random mixture itself is conceived of as a system of fortuitous combinations manifesting themselves by different possible associations of the mixed elements. In the second place, we have noted above, how the subject represents to himself the action of drawing the counters as tied to the quantitative values of the categories of elements in play; and thus as soon as he accepts the idea of random mixture he begins to quantify the probabilities themselves. In this way, a sort of operative scheme of how to anticipate intervenes in both cases, and we will find that the role of this scheme enters into the construction of the operation of combinations.

In stage III, finally, the intuitions cease to be global, and the refinement of the idea of chance and the calculation of probabilities are conceived of as fractions of certainty related to the totality of possible combinations. These intuitions show the achievement of the operative system itself which allows the establishment of combinations. In this way, the intuition of random mixture, which in stage II constituted the notion which was antithetical to the idea of operative "grouping" (*groupement*), ends up by being conceived of as the model for grouping itself, with the following single difference which contrasts precisely the rational idea of chance with the idea of simple determination: This difference is that only the totality (large numbers) is determined (by the complete multiplicative grouping of the possible combinations) the reasoning bearing on the fractions remaining polyvalent.

We see here the program which remains for us to fulfill: Chapter VI is going to permit us to analyze further the functioning of this quantification of probability which was announced in the course of Chapters IV and V. After that it will be a question of studying the development of combinatoric operations themselves. We have just seen that these combinatoric operations furnish us the key for the whole evolution of the notions of chance and probability.

VI. The Quantification of Probabilities[1]

In the course of the first part of this work, we have studied the reactions of the child faced with physical situations in which he is led, as in daily experience, to construct an idea of chance. In this second part of the work we have given ourselves the task of analyzing how the subjects' reactions are related to chance in an activity such as random drawings. The reactions to physical chance as well as to the actions of random drawing of lots have shown us the existence of two conditions essential for the development of probabilistic notions. On the one hand, as we have just seen, this progress of the interpretation of chance depends on the child's capacity to construct the combinatoric operations (combinations, permutations, and arrangements). On the other hand, the progress supposes the gradual ability to establish a relationship between the individual cases and the whole distribution; this ability to establish relationships requires logical and arithmetical operations in general, thus an operative quantification. It is now, therefore, a question of studying the mechanism of this quantification which will assure the transition between all the preceding chapters (notably

[1] With the collaboration of Myriam van Remoortel, Gaby Ascoli, and M. A. Morf.

the last two) and the third part of this work which will be an analysis of the combinatoric operations.

The preceding chapter allowed us to pose the problem: Faced with a set of colored counters where the numbers of each color were allowed to be different, the subjects of stage I are not concerned with the quantities at play; those of stage II (from seven to eleven years of age), take quantity into account at first, but not the modifications of the number contained in the sack after each of the successive drawings; only those of stage III make precise calculations. It is worth the trouble, therefore, to try to find out in detail why the child is so late in recognizing the numerical relationships which obviously seem to determine the probabilities of what is drawn, and the mechanisms of the logical and arithmetical operations on which judgments of probability depend.

1. *Technique and general results.*

To understand the functioning of this quantification, we will no longer be satisfied, as in the preceding chapter, with studying the drawing from a single sack containing a complex mixture, but we will present the subject with two small collections, each composed of a very small number of elements so that we will be able to follow step by step and in detail the relationships established.

Thus we present the child with two sets of white counters, but with or without a cross on the back. The subject takes note each time of the exact make-up of his two sets. After this we mix them up, each set separately, and put them on the table (only the top side of each element visible), and the child is to decide in which of the two sets he has the greatest chance of finding a cross on the first try. The questions asked were the ten as follows (not necessarily each question to every subject, but in general):

1. Double impossibility: the two collections, each with two or three pieces, but none of them having a cross on the other side. The child having seen beforehand each of the sets of counters in front of him, we ask him nonetheless questions to know if there is more chance of getting a cross in one or the other of the sets.

2. Double certainty: the favorable cases, for example, with two out of two or four out of four (which we will write as 2/2 and 4/4), that is, all of the pieces have crosses. Since the collections are unequal in number, the question again is to know in which case the chance is greatest.

3. Certainty-impossibility: one of the two collections with as many crosses as pieces (e.g., 2/2) and the other the same number of pieces but no crosses (therefore, 0/2). The question is still to say which set has the best chance of turning up a cross.

4. Possibility-certainty: a collection with one favorable possibility out of two (that is 1/2) and the other, one out of one (1/1), or again 1/2 and 2/2, etc.

5. Possibility-impossibility: e.g., 1/2 and 0/2.

6. Identical composition of the two collections: e.g., 1/2 and 1/2.

7. Proportionality: one of the two collections has one cross out of two pieces and the other two crosses out of four pieces, or 1/3 and 2/6.

8. Inequality of favorable cases and equality of possible cases: e.g., 1/4 and 2/4.

9. Equality of favorable cases and inequality of possible cases: e.g., 1/2 and 1/3.

10. Inequalities respectively of favorable cases and possible cases, without proportionality: e.g., 1/2 and 2/3.

In addition, we asked questions about analogous problems, but with a single set. For instance, we put two white counters and a black one in a sack: Which is most likely to be drawn out? Questions of this second type are used to give information on certain points of relationship between the results which are going to follow and those of the preceding chapter.

The reactions of the subjects to these different questions could be classed under the three usual stages. During the first stage (four to seven years), there is the absence of any comparisons based on the quantified relationships in play (level I A) or there is an intuitive comparison deriving from the perception of striking disproportions (level I B). In the course of the second stage (seven to eleven years) a single relationship is seen: The child compares the favorable cases with the unfavorable ones, but he does not construct a relationship between the favorable cases and the possible ones; it follows then that there is still a failure to understand proportionality (level II A) or perhaps the proportions are discovered when they are easily perceived (e.g., $1/2 = 2/4$), but with no generalization (level II B). Finally, in the course of the third stage (eleven years and over) relationships are established between favorable cases and possible ones in all cases and assure the understanding of probabilities, even when the favorable and possible cases differ as in the tenth situation. The problem is solved either by a calculation with fractions, or early in this stage, by successive analyses of the different intuitive structures and their logical relationships.

2. *The first stage. Level I A: Absence of logical and arithmetical comparisons.*

In the course of stage I we will distinguish two levels, determined by whether the subject is not concerned with quantitative relationships or whether he begins to make intuitive comparisons from this point. We will begin with some examples of substage I A. But we must point out here that the principal aspects of the cases we are going to describe ought to be interpreted as a whole. The questions asked are, in fact, of such a nature that the subject himself answers haphazardly, having always one chance in two of getting a correct answer. We must not then expect generally wrong answers, but ought to compare the whole of the reactions to see if the child has understood or not. On the other hand, it goes without saying that, among the number of reasons that the child will call on, he may also cite quantities. But as we have just seen in the preceding chapter, we are unable to speak about quantification of probabilities unless the quantitative bases of the children become systematic and are not simply joined to others which contradict them, since the subjects have no general comprehension of the situation. Here are some examples:

JEA (4; 6): Question 4 (possibility-certainty), 1/3 and 3/3: "Where are you most likely to get a cross right away?" *"There*

(1/3)."Why?" *"It's easiest because there's only one cross. There (3/3), there are three of them. It's easiest when there's nothing but a single cross."* "And here (3/3 and 1/6)? Where are you most likely to get a cross on the first try?" *"Here* (1/6) *because there's only one."*

LIL (4; 6): Question 4 (possibility-certainty), 1/3 and 3/3: *"Here* (3/3) *because there are three of them here."*

Question 8 (inequality of favorable), 1/3 and 2/3: *"Here* (2/3) *because there are two of them."*

Question 6 (identical make-up), 1/3 and 1/3: He chooses the group on the right *"because there is only a single one."* "And that one?" *"Only one there too."* "And so? . . ."

Question 9 (equality of favorable), 1/3 and 1/4: *"It's the same thing because there's just one here and one there too."* "And with that situation (1/6 and 1/3) with which are you most certain?" *"That one* (1/3) *because there's just one."* "And with these (1/3 and 1/6), where are you most certain to get a cross the first try?" *"Here* (1/6) *because there are more."* "But how many crosses are there?" *"One."* "Then with which are you most certain?" *"Here* (1/6)." "Why?" *"Because there are more."*

Question 7 (proportionality), 1/2 and 2/4: *"Here* (2/4) *because there are two."*

Lil seems, therefore, to solve correctly problems 4 and 8, but when his answers are compared with the following ones, we see that he decides simply by an examination of the favorable cases without taking into account the possible ones—the answers to question 9 give evidence of this.

BUR (4; 8): Question 9 (equality of favorable), 1/2 and 1/3: "With which one are you most certain?" *"There* (1/2)." "Why?" *"I don't know."* "Take a try with the one where you are most likely to get one on the first try." (He chooses from the 1/3 group.)

Question 2 (double certainty): "With which pile are you most certain?" *"There and there* (correct)." "Why?" *"Because there are two of them."* "Is it not more certain with one than with the other?" *"Yes, there* (on the right), *because there are two."*

FRA (4; 9): Question 9 (equality of favorable), 1/4 and 1/3: *"There* (1/3)." "Why?" *"Because there's one cross."* "And with these (1/2 and 1/3)?" *"There* (1/3)." "Why?" *"Because . . .*

no, there (1/2), *because I've already chosen the other too much."*

Question 4 (possibility-certainty), 1/2 and 1/1: *"There* (1/2)." "Why?" *"Because."*

Question 2 (double certainty), 2/2 and 2/2: *"There* (on the left) *because there are two of them."*

Ros (5; 2): Question 9 (equality of favorable), 1/2 and 1/3: *There* (1/3)." "Why? . . ."

Question 4 (possibility-certainty), 1/2 and 2/2: *"There* (2/2) *because there are two crosses."* "And how about that (2/2 and 2/3)?" *"There* (2/3)." "Why?" (He hesitates and is ready to try the other one and then comes back to 2/3.) "And that (2/3 and 2/2)?" *"Here* (2/2) *because there are two of them."* "And with this (2/7 and 1/1), with which one are you most likely to get a cross on the first try?" *"Here* (2/7) *because there are two of them."*

Cab (5; 3): Question 9 (equality of favorable), 1/3 and 1/2: "If on the first try you want to get a counter with a cross, which is best?" *"Here* (1/2)." "Why?"

Question 7 (proportionality), 1/2 and 2/4: *"Here* (1/2) *because there's only one."* "And how many are there in the bigger pile?" *"Two."* "Then why did you choose here (1/2)?" *"It's easier in the little one."* (See question 9.)

Question 8 (inequality of favorable), 1/4 and 2/4: *"Here* (1/4) *because there's only one."*

Question 4 (possibility-certainty), 1/2 and 2/2: "Where will you choose to get one on the first try?" *"Here* (1/2)." "How many crosses are there in the two piles?" *"Two there and one there."* "Then where is it easiest to find the cross?" *"Easiest here* (1/2)." "Why?" *"Because there's just a single one."* "Are we more likely to get one here (2/2) or not?" *"Certain."* "And there (1/2)?" *"Certain."* "Why?" *"There is one there."* "Where do you want to try?" *"There* (1/2)."

Question 2 (double certainty), 2/2 and 2/2: "Where are we most sure?" (He points to the one on the left.) "Why?" *"There are two of them."* "And there?" *"Two."* "Why do you choose this one then?" *"Because there are two."*

Question 1 (double impossibility), 0/2 and 0/2: (He shrugs

his shoulders.) *"You can't win."* "Why not?" *"There aren't any crosses."*

MER (5; 3): Question 2 (double certainty), 2/2 and 2/2: "Do you remember?" *"One cross there and another there; one here and another here."* "Where is it better to choose?" *"There on the right."* "Why?" *"Because there's a cross there."* "But there are crosses all over. Why do you choose there?" *"Because they are all full."*

Question 1 (double impossibility): *"There aren't any."*

REH (5; 7): Same reactions as Mer.

WEI (5; 9): Question 9 (equality of favorable), 1/2 and 1/3: *"Here* (1/2) *because there are only two counters."*

Question 7 (proportionality), 1/2 and 2/4: *"Here* (1/2) *because there are two counters and one cross* (i.e., a smaller set)."

Question 8 (inequality of favorable) 1/4 and 2/4: *"There* (1/4) *because there's only one cross."*

GAT (6; 2). We establish a control using a single set having a red counter and two white ones in a sack: "If you draw out a single counter without looking, what are you most likely to get, a red or a white?" *"A red one."* "And with this grouping (a red and three white ones)?" *"A red one."* "And this way (a red and four whites)?" *"A white."* "Why? . . ."

Question 8 (equality of favorable), 1/3 and 1/2: "In which group are you most likely to get a cross?" *"Here* (1/3) *because there is one here."* "And there (1/2)?" *"Same."* "Then it is not easier to get one here (1/2)?" *"No. Here* (1/3)."

We see that it is not an exaggeration to characterize this level I A by the absence of any ability to make quantitative comparisons between the two sets where it is a question of deciding in which set the probability is greater for getting a counter with a cross. Since the child at this level has neither elementary logical operations (inclusion of a part in a whole where conservation is possible) nor arithmetical operations formative of the series of whole numbers, there is nothing surprising in this. But, in addition to this lack of logical and mathematical operations, another problem arises which precisely these observations can clear up for us: It is a fact that the numbers used here are exclusively intuitive (or figural) ones, one to four or five, and that an elementary probabilistic intuition could be brought out with such small sets. The absence of such intuitions, therefore, is of great interest and even allows us to hope for some explanation of an operative order for all the facts observed up to now in the beginning of stage I.

Let us note, first of all, that leaving aside questions 1, 3, and 5 (different impossibilities) which serve as controls, none of the questions 2, 4, and 6 to 10 yield systematically correct answers. Any correct answers are due to chance (as we see by a comparison with all of the other answers given by each subject), or, somewhat rarely, to a momentary intuition, but which will really come into being only in level I B or level II in general.

Question 4 (possibility-certainty) brings out clearly a first essential factor of prediction proper to this level. For example, in the presence of three crosses out of three pieces in one of the sets, and a single cross out of three pieces in the other set, the child ordinarily considers (cf. Jea, Fra, and Cab) that he is more certain when looking in the set 1/3 because it is a matter of finding only one cross. In the set 1/3 *"It's easier because there is just a single cross. There* (3/3) *there are three of them. It is easier when there is just a single cross* (Jea)." It is true that other subjects, like Lil and Ros, first choose from the groups 2/2 or 3/3, but after some questioning (particularly with Ros) we see that there is no question of probabilistic reasoning.

Question 8 (inequality of favorable cases, for example, 1/4 and 2/4) yields analogous remarks: Cab chooses the set 1/4 because the cross is found there all alone (as different from 2/4 where there are two of them), while Lil seems to answer correctly, but without the rest of her remarks confirming the existence of a coherent probabilistic judgment.

Question 9 (equality of favorable cases and inequality of possible cases, for example 1/2 and 1/3) yields in its turn a clearly negative reaction. If the child had the least bit of quantified probabilistic intuition, it would be easy for him, even without counting, to understand that a counter with a cross is more easily gotten from a smaller set than from a larger one. While they often prefer the smaller set when working with question 7 (proportionality) where the size of the set actually does not at all matter (cf. Cab and Wei), the subjects of level I A answer question 9 with no awareness of the quantitative relationship. They decide that these probabilities are the same (since the favorable cases are so, unlike the possible ones) or choose paradoxically the larger group, or even choose simply haphazardly. Thus Lil makes the judgment that the chance of one out of three is the same as one out of four, then with 1/3 and 1/6, chooses 1/3 *"because there is only one* (the same reasoning as Jea for question 4)," and later chooses 1/6 *"because there are more."* Bur, Ros, and Cab choose at random. Fre does the same and then reverses his choice *"because I already chose a lot in the other,"* as if probability were absolutely indeterminate. Wei seems to answer correctly, but later in the questioning contradicts the existence of a systematic intuition. Gat finally reacts as if simply his desire was going to assure the success of his choice.

In these conditions, question 10 (both favorable and unfavorable cases unequal) is beyond the level of the subject, as is true with question 7 (proportionality) also. Nonetheless we asked this last question, and the answers confirmed our earlier results. Either the child chooses the largest group because it contains the largest number (in the absolute sense) of favorable cases (Lil, 2/4 *"because there are two of them"*), or he chooses the smallest (Cab and Wei) precisely because there is a single one. *"Because there is only a single one,"* says Cab in answer to question 8 (inequality of favorable cases).

These different reactions, therefore, exclude any searching for a logical or arithmetical or quantitative comparison between the two sets from the point of view of a relationship between the favorable and the possible cases. Said in a simpler way, the child is just not at all concerned with the possible cases and considers only the number of favorable cases as if that were an absolute value. He is so little concerned with the possible cases, as such, that in the situation described by question 2 (double certainty) and 6 (identical composition, that is, equality of favorable and possible cases), the child does not make (except Bur at the beginning) any comparison in general of the two identical groups and chooses one arbitrarily.

How then are we to explain the whole of these facts? In addition to all the detailed factors that we know already, two fundamental characteristics of an operative order (and, moreover, closely related) seem to dominate the initial reactions of the subjects with regard to chance and probability: On the one hand, there is the difficulty of handling the disjunction and, on the other hand (which is about the same but from a different point of view), the absence of a linking of the parts into stable wholes.

For example, let a mixed set B be made up of red counters (this first subset we call A) and of two white counters (this complementary subset of A is called A′). We can then unite in any order the parts A and A′ of the whole set B by means of the operation of union ($A \cup A' = B$) or by using the relative complement operation ($B - A = A'$ or $B - A' = A$). It is the reversibility of this operation which assures the conservation of the total, B, which is either the result of the union ($A \cup A' = B$), or continues to act as a reference set with the relative complements ($B - A = A'$ or $B - A' = A$).

In addition, let us call p the proposition "x is a member of A"; p' the proposition "x is a member of A′"; and, finally, q the proposition "x is a member of B." If set A is part of B (inclusion) and set A′ is also (complementary inclusion), then "x is a member of A" implies "x is a member of B" and "x is a member of A′" implies "x is a member of B," that is, if p then q and if p' then q." Suppose that after the counters are mixed in a sack, we know about an x only that it is in B, without knowing if it is in A or in A′. We will have in this case: "if q then p or p'," that is, "if x is a member of B, then it is in A *or* in A′." It is this interpropositional operation (p or p') corresponding to the union of sets ($A \cup A'$) that we call disjunction.

The psychological problem brought up by the reactions at level I A is thus to determine whether or not the child understands the operations of union of complementary sets ($A \cup A' = B$) and of disjunction of propositions ("if q then p or p'"), which seem to constitute the condition that is needed before there is any logical development and quantification of probability.

Let us start then from a unique set, for example, of a single red counter (A) and two white ones (A′) and ask ourselves how the child reasons from the point of view of disjunction (developing only afterward the union of sets). For him to understand that drawing a white counter is more probable than drawing a red one, it seems that it would be necessary to assume the two following operations: Since the whole set contains two subsets (red and whites), I can draw either a red or a white (disjunction); but since there are two white ones and only one red (comparison of quantities), the two sides of the alternative are not equal and it is more likely that I will draw a white one. The fact is that the opposite happens as if the child cared not at all about the unions at play nor the disjunction and simply said to himself, *"They are asking me for a single piece, a red one, and since there is exactly one red one, I will bet that I get it and not bother about the white ones* (see the case of Gat who reasons this way until there are four white ones)." If we now have the child compare two sets, the one containing three crosses out of three pieces, and the other a single cross in six pieces, he will say to himself as Jea did that, to get a cross on the first drawing, we must choose the second group since there is exactly one there and not three! In other words there is no disjunction because the subject simply does not bother himself about the different possible cases, that is, about the totality as such or about the alternatives which it offers. He goes straight to the favorable case, neglecting the others.

Such a reaction makes one think immediately of the well-known facts about understandings based on either/or propositions: With these the children never choose—either they opt for both terms at once, or they think of the one and forget the other. For example, faced with a picture representing a dog or a cat, a subject (2; 10) says that it is a dog *"because it's gray,"* [1] as if the fact of being a gray quadruped did not imply precisely the disjunction "either a dog or a cat" (or even several other possibilities).

The difficulty of working with disjunctions (a very general difficulty which goes far beyond the quetsion of chance) shows the existence of an essential trait of the thought of the child and as well something of great importance concerning the intuitions of probability: the inability of the child to reason on the possible. He does not choose between several possibilities, but goes directly to

[1] See *La formation des symboles chez l'enfant* (Paris: Delachaux and Niestle), p. 246.

one of them which he considers as the only real one (or only desirable one, that is, achievable), one which he does not understand as one possibility among others but as a privileged one which eliminates all the others. This absence of disjunctive reasoning then ends up, in fact, at an inability to understand even the notion of possibility, since this notion necessarily implies the consideration of several possibilities, and is then, one might say, disjunctive by its very nature. When the child chooses the set in which the wanted element exists alone (among different ones), we can consequently interpret his reaction by saying not only that he fails to reason by disjunctive comparisons but even that he does not understand the very notion of possible cases and of possibility itself. He transforms incorrectly a possible case—that is, relative to other possible cases—into a certain or quasi-certain case—that is, a case envisaged absolutely as a unique situation and independently of all the possibilities. The absence of disjunctive reasoning from a formal point of view, the feeble differentiation between the possible, the real, and the necessary from the modal point of view, and the incomprehension of the fortuitous and the probable are three closely linked phenomena.

This first aspect of the initial reactions of our subjects derives from an aspect of their thought which is even more profound and general, and that is the difficulty of forming the union of the parts into a stable whole (hence the inability to move disjunctively from this whole to the parts).

Even though the child lacks these modal notions of possibility and necessity, he could understand in fact that a certain counter is easier to find when there are only two counters than with three or four, for the simple reason that the first is a greater proportion of the whole. And, in fact, the child does have occasional recourse to arguments which seem to make elementary relationships of the part to the whole: Lil, for example, comparing 1/3 with 1/6 chooses 1/3 "*because there is only one* (cross)" rather than 1/6 "*because there are more of them* (counters with or without crosses)." It happens, however, that the reasons the child invokes in such cases show precisely a systematic difficulty of forming the union of the parts of the whole, or, if you like, in including the favorable cases (crosses) in the whole of possible cases (counters with or without crosses) seen as a totality. In other words, it is not only because it is a question of possible cases, that is, being able to draw only the ones or the others, that the child does not think of the whole when he thinks of the favorable cases. It is also, and even primarily, because he does not know how to construct a reversible union (relative complement) of the part with the whole. Even more, as we saw above, it is precisely because he lacks such understanding of union that the child does not come to the operation of disjunction, nor, consequently, to the notion of the possible (it can be either this, or those).

In fact, when the child seems to refer to relationships of the part with the whole, we notice very quickly that he is in reality invoking only one of the two numbers available, either the one of the part (favorable cases), or of the whole (the set of all counters, but without disjunction precisely because he lacks the ability to make a relationship between the part and the whole, thus the situation is not conceived of as all the possible cases). Lil, for example, who at first seems to reason by relationships of the part to the whole (he prefers 3/3 to 2/3 and 2/3 to 1/2) shows in his answer to question 9 that this is not so at all: 1/3 seemed to him the same as 1/4 *"because there is one here and one there"*; and sometimes 1/3 is chosen over 1/6 and sometimes the opposite because Lil is thinking of the part in the first case and of the whole in the second. He makes no relationship between the two cases in either choice. Ros decides that the probability with two out of seven is greater than the certainty of one out of one because *"there are two crosses"* in the first set. Cab, on the contrary, faced with the proportions 1/2 and 2/4 believes that success is *"more likely in the smaller set,"* and so on. In brief, the child takes into consideration sometimes the part, sometimes the whole, but never the relationship of union which unites them.

It is easy to understand the causes of this failure as well as its effects on disjunction and on the modality of judgment. Its cause is the irreversibility of the thought of the child which we emphasized in Chapter I. Since they are unable to imagine, by any representation, the two directions of the action, they do not grasp the necessary reciprocity of the operations of union ($A \cup A' = B$) and of relative complement ($B - A = A'$ and $B - A' = A$): A part (for example, A) once dissociated by thought from its complement (A′), the whole (B) is not conserved as a possible union of the two. For this reason, when the child moves his attention to one part of the set of counters, say, the ones with crosses, he neglects the whole and fails, consequently, to establish any relationship between the part and the whole. On the other hand, we have seen in Chapter I that it is this same general lack of operative reversibility which keeps the children of stage I from understanding the irreversibility which is proper to random mixture and consequently to the notion of chance. Finally, it is clear that the absence of the operation of reversible union explains the child's inability to handle disjunction and thus the lack of differentiation of the possible, the real, and the necessary. Because of his failure to keep in mind the whole (B) as a permanent union of the parts ($A \cup A'$), the subject will not be able to conceive as constant evidence that if a counter is in B, and if the only thing known about it is that it is in B, then it is either A or A′, that is, that the whole (B) is a union of all the possible cases.

3. *The first stage. Level I B: Reactions intermediary between stages I and II.*

As we have already noted previously, the child could conceive chance, that is the incomposable and irreversible, only by reference to composable and reversible operations. We can now add that, when such operations are still lacking, not only does any idea of chance remain inaccessible, but even more so, any notion of probability, since such notions must include understanding the relationships of favorable with all possible cases.

In the preceding cases, Gat expected to find a red counter mixed with two or three white ones more easily than the white ones, but when a fourth white one was added, he then reversed his decision. At a certain point the increasing number of white counters brings him to the decision that one of the white counters would be drawn more easily than the single red one. It is these kinds of intuitive rules which in their development will characterize level I B.

Eis (6; 1): Question 9 (equality of favorable), 1/2 and 1/3: *"Here* (1/2) *because I'll get a cross right away."* "Why?" *"Because there are fewer."* "And with these (1/5 and 1/6)?" *"Here* (1/5) *because there are fewer."*

A single group (one red and two whites): "What color are you most likely to get if you take a counter?" *"A red one."* "Why?" *"I'll get it because there's only one of them."*

Question 7 (proportionality), 1/2 and 2/4: *"Here* (1/2) *because there are two."*

Question 10 (both unequal), 3/4 and 1/2: *"Here* (1/2) *because there are fewer."*

Question 8 (inequality of favorable), 1/4 and 2/4: *"Here* (2/4) *because there are two crosses."* "Why is that more likely?" *"Because there are two with crosses and two without."*

Question 4 (certainty-possibility), 2/2 and 1/2: *"Here* (2/2)." "Why not here (1/2)?" *"Because there is one without a cross."*

Bru (6; 2): Question 9 (equality of favorable), 1/2 and 1/3: *"Here* (1/3) *because there's one cross."* "And there (1/2)?" *"One there too."* "With which one is there a better chance?" *"Here* (1/3). *Oh, no, here (1/2) because there are two."*

A single group (one red and two whites): "What will you most likely get if you take one?" *"The red."* "Why?" *"Oh, no, a white one because there are more of them."*

Question 8 (inequality of favorable), 1/4 and 2/4: *"Here* (2/4), *there are more crosses."*

Question 10 (both unequal), 1/2 and 2/5: "Where do you have the best chance?" *"Here* (2/5)." "Why?" *"Because there are two."* "And this one?" *"No, here* (1/2) *because there is one with a cross and one without, and that's all there are."*

Question 7 (proportionality), 1/2 and 2/4: *"More chances here* (2/4) *because there are more crosses."*

Question 2 (double certainty), 2/2 and 2/2: *"There* (the left one) *because there are two crosses."* "And this one?" *"That one too. It's the same thing, easy."*

Sci (6; 6): Question 9 (equality of favorable), 1/2 and 1/3: *"Here* (1/2) *because there are fewer of them."* "Of which?" *"Counters."* "And?" *"It's easier."*

A single group (one red and two whites): "If you take just one, what color will it most likely be?" *"Red."* "Why? . . ." (we repeat the question). *"Red."* "And this way (two reds and a white)?" *"A red one. . . . No, the white one."*

Question 6 (identity), 1/2 and 1/2: *"Same."*

Question 7 (proportionality), 1/2 and 2/4: *"Here* (2/4) *because there are four of them."*

Bov (6; 7): Question 9 (equality of favorable), 1/3 and 1/4: *"Here* (1/3) *because there are three."*

Question 6 (identity), 1/6 and 1/6: *"Either one."*

Question 7 (proportionality), 1/3 and 2/6: *"Here* (1/3) *because there are three of them."*

Question 8 (inequality of favorable), 1/4 and 2/4: *"Either."* "Why?" *"Because there are four in both piles."*

Question 2 (double certainty), 2/2 and 2/2: *"Either one."*

Single group (one red and two whites): (Silence; no decision.)

CHAN (7; 6): Question 9 (equality of favorable), 1/2 and 1/3: *"Here* (1/2) *because it's easier."* "And how about this way (1/2 and 1/5)?" *"There* (1/5)." "Why?" *"Because there's a cross. No, here* (1/2) *because there are fewer."*

Question 8 (inequality of favorable), 1/4 and 2/4: *"There* (2/4) *because there are two crosses."*

Question 7 (proportionality), 1/2 and 2/4: *"There* (1/2), *no, here* (2/4) *because there are two crosses. No, it's easier here* (1/2)." "And this way (1/3 and 2/6)?" *"Here* (1/3), *there's a better chance."*

Question 9 (equality of favorable), 1/4 and 1/5: *"There* (1/4) *because there are fewer."*

It is proper to notice two different categories in the problems asked: There are questions which put two variables into play at the same time—questions 7, proportionality, and 10, in which both the favorable cases and the possible cases vary from one group to the other. These two questions will not be resolved, at least not systematically, until level III. All of the other questions contain only a single variable and thus seem to be all of about equal difficulty. Now the difference between stage II and level I B is that in stage II all the questions with a single variable will be systematically resolved while at level I B, some are and some are not. The problem then is to understand why there is this distribution of correct answers which are, in appearance, arbitrary.

When questions of the same logical difficulty yield solutions which are not general, it is usually because the solutions are not yet based on operative reasoning, but are simply intuitive. Thanks to perceptive or imagined schemes which are sufficiently effective when the situation is simple, the child understands; but these solutions are not susceptible to generalization when the situation gets more complex because there is a lack of logical operations.

When considering each particular case, it is easy to see that this explanation is correct. The child has difficulty in arriving at a means of linking the favorable cases, as part A, with the total possible cases B, and then of constructing a detailed (numerical) quantitative comparison between the part A and the whole B or between part A and its complement A′. Now, without yet being able to construct these links and this quantification by means of effective operations (that is, mobile and reversible operations), the subjects at this level I B arrive at an intuition which marks a definite progress over level I A: The more counters there are, the more possibilities there are also, and thus the less chance there will be for getting the desired counter on the first draw. This is, however, only an intuition, that is, it is based essentially on a perceptive or imagined configuration of the set, and not yet on unions and disjunctions which could be generalized for all cases.

Among others, there is one striking fact in this regard: With only few exceptions, the question about a single group (i.e., Do we have a better chance of getting a red or a white when drawing from a group with two white ones and one red?) is more difficult than the analogous one asked about two groups (Do we have a better chance of drawing a cross from a group of two when one has a cross, or a cross from a group of three?). Logically the two problems are equivalent. From the point of view of representative intuition, however, we see immediately the difference between them: When two groups are present, the single fact of comparing the one with the other makes it easier to disengage the relationship of the part with the whole since the relationship in the first group is different from the one in the second. In the case of the single group, on the contrary, the union of the elements in a single intuitive whole hinders the subject from thinking simultaneously of the whole and the parts. In fact, faced with two groups with a single variable such as 1/2 and 1/3 or 1/4 and 2/4, the child notices immediately the difference between two and three in the first case or one and two in the second. It is this difference alone then which determines his judgment: There are more unfavorable cases in 1/3 than in 1/2 and more favorable cases in 2/4 than in 1/4. But on the other hand, in the case of a single group composed of 1/3 (one red of three counters), to compare the favorable case (1) with the unfavorable ones (2), he must see both of them as functions of the whole. That is, he must compare 1/3 with 2/3, which calls for a greater abstraction since the whole is single: Comparing then a part (1) only with another (2) without seeing that each element has an equal chance of being drawn (from failure to tie them together as one whole), he falls into the idea that in drawing a single counter, he will draw the red one because it is single.

We understand then the reason for some of the errors persisting at this level. When comparing two groups, the subject thinks of the invariable term (of the favorable cases, that is, the part; or of the possible cases, that is, the whole) instead of fixing his attention on the variable, and is fooled every time. Thus Bov judges the chances with 1/4 and 2/4 as the same because he thinks of the group of counters (4), while Bru chooses 1/3 over 1/2 and Chan 1/5 over 1/2 because they are thinking of the single cross. In these particular cases, as in the one with the single set, the subject shows that he does not yet proceed, in fact, by linking the favorable cases (the part) as a set of possible cases (the whole), that is by systematic disjunctions. He limits himself to two intuitive anticipations: When he has an equal number of possible cases, he centers his attention on the favorable ones and chooses the group with the larger number. And when he has an equal number of favorable cases, he thinks about the unfavorable ones and chooses the group with the smaller number. This is the limit of progress found in level I B. But compared with the reactions at the first level, I A, it is seen as meaningful and is a preparation for the operative inclusions and disjunctions of stage II.

4. *The second stage: The general success of comparisons with a single variable. Level II A: Systematic failure with questions of proportion.*

As soon as the first logical arithmetical operations are acquired, that is, at about seven years of age, the child no longer experiences difficulties in linking the favorable cases (part A) with the unfavorable ones (A′) making a whole (B) which is seen as an assemblage of the possible cases and allowing varied disjunctions as functions of the sets A and A′. The questions bearing on two variables which assume a notion of proportion require, on the contrary, more complex operations which appear only in level III, although they are prepared for in the course of substage II B.

Here are some examples of level II A (beginning with an intermediate subject, all of whose answers are correct but who reasons as does a subject of level I B):

GIR (6; 6): Question 9 (equality of favorable), 1/2 and 1/3: *"Here* (1/2) *because there are two."* "And what about this (1/5 and 1/6)?" *"There* (1/5) *because there are five, and not there* (1/6) *because there are six."*

Question 4 (certainty-possibility), 2/2 and 1/2: *"That one* (2/2) *because there are two; yes, that's certain* (shrug of the shoulders)."

Question 8 (inequality of favorable), 1/4 and 2/4: *"That one* (2/4) *is more certain because there are two."*

Question 10 (both unequal), 1/2 and 2/5: *"Here* (1/2) *because there are two."*

Question 7 (proportionality), 1/2 and 2/4: *"Here* (2/4) *because there are two of them."* "Isn't it just as good in both piles?" *"Yes, there are two crosses here, and there one* (feeling of proportionality)." "And this way (1/3 and 2/6)?" *"Here* (1/3) *because there are three."* "Same situation with both?" *"No."*

PIL (7; 8): Question 9 (equality of favorable), 1/2 and 1/3: *"Here 1/2 because there is one cross."* "And here (1/3)?" *"There are two of them without crosses."* "Then why is it better with this one (1/2)?" *"Because there are two of them in all."*

A single collection (one red and two whites): *"We are more likely to get a white one because there are two of them."* "And with this arrangement (two of each)?" *"Both sides are the same."*

Question 10 (both unequal), 1/2 and 2/3: *"Here* (1/2) *is more certain because there is only one without a cross while in the other there are three* (in all) *and then I can choose from there and not get any."*

Question 2 (double certainty): *"We can get a cross out of both right away."*

Sou (8; 3): Question 8 (inequality of favorable), 1/3 and 2/3: *"Here* (2/3) *is more likely because there are two crosses. That's more than one. I can get it twice here* (2/3) *and only once there* (1/3)." "And how about this (4/5 and 3/5)?" *"Here* (4/5) *because there are four with crosses, and only three there."*

Question 9 (equality of favorable), 1/5 and 1/4: *"Here* (1/4) *it's more certain because there are only four counters. I'd lose fewer times than with five."*

Question 7 (proportionality), 1/2 and 2/4: *"Here* (1/2) *because if I don't get it, I'll lose only once."* "And with this (1/4 and 2/8)?" *"Here* (1/4) *because I'm certain to lose fewer times."* "How many?" *"Three times, and in the other six times."* "How many times can you win with 1/4?" *"Once."* "And in the other?" *"Twice."* "And so?" *"I prefer this one then* (1/4)." "And with this (2/6 and 4/12)?" *"Here* (4/12), *I'll win more times, but I'll also lose more times. But I prefer it* (4/12) *because I'll win more often."* "And in the other?" *"I'll win twice and lose four times."* "And the first one?" *"I'll win four times and lose eight times."* "And so? . . ."

Question 10 (both unequal), 1/3 and 2/3: *"Here* (2/3) *it's more certain because I can win twice and lose only once. There* (1/2) *I'd win one time and lose one time too."* "And with this (4/9 and 2/5)?" *"Here* (2/5) *because I'll lose less."*

Adl (8; 6): Question 8 (inequality of favorable), 1/3 and 2/3: *"Here* (2/3) *because there are two with crosses."*

Question 9 (equality of favorable), 2/3 and 2/4: *"Here* (2/4) *there are two pieces without crosses. Here* (2/3) *it's easier because there is just a single one without a cross."*

Question 7 (proportionality), 1/2 and 2/4: *"There* (1/2) *it's easier because there's only one of them without a cross."*

Question 10 (both unequal), 2/3 and 3/4: *"More likely with this one* (3/4) *because there are three crosses, and there* (2/3) *only two."* "And how many without crosses?" *"One and one."* "And then?" *"In this one* (2/3). *No; it's the same."*

BAY (9; 3): Question 8 (inequality of favorable), 1/3 and 2/3: *"There* (2/3) *because there are more crosses."*

Question 7 (proportionality), 1/2 and 2/4: *"Here* (2/4) *because there are more crosses. And anyway it's easier."* "And this (1/2 and 3/6)?" *"Here* (3/6) *because there are more crosses."*

Question 9 (equality of favorable), 1/2 and 1/3: *"Here* (1/3), *not there* (1/2) *because here* (1/3) *there are many without crosses."*

Question 10 (both unequal), 1/2 and 2/5 (long hesitation): *"Maybe* (1/2)." "Why? . . . (unable to answer)."

ALL (9; 6): Question 8 (inequality of favorable), 1/3 and 2/3: *"There* (2/3) *because there are two of them."*

Question 9 (equality of favorable), 1/4 and 1/5: *"Here* (1/4) *because there are three without crosses, and there* (1/5) *four."*

Single group (one red and two whites): *"We are more certain to draw a white since there are two of them."*

Question 10 (both unequal), 1/2 and 2/5: *"Here* (2/5) *because there are two."*

LYO (10; 7): Single group: *"A white because there are two."*

Question 7 (proportionality), 1/2 and 2/4: *"Here* (1/2) *because there are only two."*

The progress seen by comparison with level I B is clear. All the questions with a single variable are solved immediately, including the question about the single group (Pil, All, and Lyo). Moreover, the argumentation of the children takes into account simultaneously the favorable cases (A), the unfavorable cases (A′), and the totality of possible cases (B). Thus each of the subjects at this stage solves simultaneously questions 8 and 9 (inequality either of the favorable or the unfavorable cases) while in level **I B** one of these questions could be solved and the other not. The arguments, therefore, in the one case are seen to be the same as those used in the other. For example, Sou prefers 2/3 to 1/3 because he *"can win twice with this* (2/3)" while he chooses 1/4 instead of 1/5 *"because there are four counters and I can lose less than with five."* There is no question that we have here a relationship established between A and A′ (unfavorable cases) and with the whole B (the four or five counters). In addition, the reasoning on the questions with two variables, although not correct because of the lack of proportionality, shows this same preoccupation. Thus Sou already understand why the relationship 2/3 is better than 1/2: *"This one* (2/3); *I will win twice and lose only once. There* (1/2) *I'll win once and lose once."*

This ability to break the whole into correlative parts according to a reversible scheme (of unions) leads immediately to the double discovery of disjunction (an element of B can be in A or in A′) and multiple possibilities: One *"can win"* and *"can lose,"* as Sou says. In fact, all the subjects of this stage distinguish two types of modalities which remained undifferentiated from the true situation in the course of stage I (when the possible was necessary). The language of the child of stage II already gives clear witness of this: *"I can. It's quite certain* (Gir)," and so on.

Still remaining is to know why the child, when he has these new operative instruments, cannot solve the questions with two variables, that is, those with proportional probabilities (question 7) or the even more complicated case (question 10). There are two related reasons for this: The first is that the questions with a single variable can be solved by simple additions or subtractions with a relationship made between the results. For example, in question 8 the whole (B) remains constant: We have then $B - A_1 = A'_1$; $B - A_2 = A'_2$ and the question is then brought back to the comparison of relative sizes between A_1 and A_2 and between A'_1 and A'_2. In question 9 it is the part A which remains constant: Then we have simply to compare B_1 with B_2 or A'_1 with A'_2. If, on the other hand, the child seeks to establish the relationships A_1/B_1 or A_2/B_2, he can be satisfied by using the logical multiplication of the relationships, without recourse to fractions, since either A or B is constant. The questions of proportionality, on the contrary, assume a double relationship and thus a mathematical operation either numerical (fraction) or extensive (geometric proportion) which is naturally more difficult. In fact, at this point it is a question of reasoning according to two systems at once, which assumes formal operations.

From this comes the second reason. We have earlier seen that questions relating to the law of large numbers were understood only in stage III because they implied, in particular, working with proportions. Regularity increases with large numbers, but it is a question of relative regularity (differences or gaps which are in proportion to the size of the numbers in play) and not absolute. What we find here is an analogous problem: The subject must understand that the number of chances is relative to the order of the size of the whole set of possible cases and no longer has an absolute value. If one can say it then, there are relative probabilities which are no longer comparable in a simple way because they lack a common element (A, A′, or B). Again then we have two distinct systems which must be united with each other, each having its own possibilities and probabilities. The difficulty is the same as with large numbers, since it goes beyond the domain of concrete operations to assume hypothetical-deductive reasoning.

5. *The second stage. Level II B: Progressive empirical solution of questions of proportionality.*

Before being able to yield a systematic solution, the questions with two variables are solved by little by little by means of empirical gropings:

RAM (9; 9): Question 9 (equality of favorable), 1/3 and 1/4: *"Here* (1/3) *we can get the cross better because there are only two without crosses."*

Questions 7 (proportionality), 1/2 and 2/4: *"Here* (2/4). *No, here* (1/2) *because there is only one without a cross."* "Why did you say the other?" *"Because there were two crosses."* "Then are two crosses better than one cross?" *"One without a cross, two with . . . it's the same because there are two crosses and two without crosses* (in 2/4)." "Then is one of the two groups better?" *"No, they are the same, but we are more certain here* (2/4) *because there are two crosses* (very perplexed)."

KEL (10; 2): Question 7 (proportionality), 1/2 and 2/4: *"Here* (1/2) *because there are fewer. No, this one* (2/4) *because there are two crosses."* "Is it the same?" *"No, this one* (1/2) *is more certain anyway. . . . No, both because here* (2/4) *there are two of each, and here* (1/2), *one of each."* "Very good. Now explain it." *"Because there* (2/4) *there are two of each, and here one of each; and then we can get a cross but each time we can also get one without a cross."* (In other words, in the relationship of either 1/2 or 2/4, for each counter with a cross there is also one without.)

Question 10 (both unequal), 2/5 and 1/3: *"Here* (1/3) *it's more certain because there are fewer."* "But there (2/5) are there more crosses?" *"Yes, but also more without crosses."*

Question 7 (proportionality), 1/3 and 2/6: We arrange the group of 2/6 into two piles of 1/3: *"In either one of the two* (1/3 or 2/6) *because they are the same. In each group, for one with a cross there are two without."* "And now what if I take away one without a cross (which gives us the situation of question 10, 2/5 and 1/3), where is there a better chance of getting a cross?" *"Here now* (1/3) *where there are fewer."* "And this way (1/3 and 2/6)?" *"Here* (1/3) *where there are fewer without crosses."* "And how many with crosses?" *"Oh, it's the same."* "And now (2/5 and 1/3)?" *"It's more difficult here* (2/5)." (Again he is fooled.) "And like this (2/6 and 1/3)?" *"In the smaller group it's better because there are fewer pieces."* (We put the pile 2/6 into two piles of 1/3.) *"It's the same."* "And if we mix them (2/6 in a single group)?" *"It's more likely in this pile* (1/3)." "But did you tell me that it is the same?" *"Yes, it is the same. Now it's the same."* "Why?" *"Here* (2/6) *there are more with crosses and more without, and here* (1/3)*fewer of both."*

MEY (11; 1): Question 7 (proportionality), 1/2 and 2/4: *"Here* (2/4) *it's more certain because there are two crosses."* "But are there two without crosses also?" *"Well then it's the same."* "Which do you choose?" *"This one even so* (1/2) *because there's only one without a cross."* "But here (2/4) did you tell me that there are two crosses? Is it the same?" *"No. I think even so that it is there* (2/4) *because there are two crosses and two without crosses, and there* (1/2), *one and one, and we can just as well get one with a cross as one without.*" "Then it is the same?" "*Yes, because where there are four counters, we can just as well get two which don't have crosses.*"

Question 10 (both unequal), 2/5 and 1/3: "*It's the same because there* (2/5) *there are three without crosses and two with crosses, and here* (1/3) *one with the cross and two without.*" "And if I put this (1/3 and 2/6)?" "*I prefer this* (1/3)." "Watch." (We put 2/6 into two piles of 1/3.) "Is it the same?" "*Yes.*" "And if we put them back together?" "*I prefer this* (1/3) *because I have a better chance of getting a cross.*" "But here (2/6) are there two crosses?" (Long hesitation.) "*Then it's the same.*" "Fine, and now how about this (1/3 and 2/5)?" "*This one* (1/3) *is more sure because there is one cross and two without.*" (We put one cross opposite two and two without crosses opposite three.) "Watch." "*Oh, there's one fewer without a cross in what remains.*" "And what if we mix them (2/5 opposite 1/3)?" "*This* (2/5) *is more likely.*"

PER (12; 2) already shows reactions proper to stage III for simple cases of proportions (question 7) such as 1/2 and 2/4. "*Both piles are the same because here there are two and two and in the other, one of each.*"

"And if we have this (1/3) on one side, and two counters with crosses on the other, how many counters without crosses would we have to add to get the same number of chances on each side?" (He gives 2/4 as equivalent to 1/3.) "*Like this we have two without crosses on each side.*" "And what if I put 1/3 and 2/6?" "*Oh, now we have the same chance.*" (In the same way he finds 2/3 is equal to 4/6 and 2/3 to 6/9.) "Why?" "*We need three times as many crosses.*"

BIS (12; 2) is even closer to stage III. Question 7 (proportionality), 1/2 and 2/4: "*In both cases, there are as many risks as there are chances. It's the same.*" "If we put this here (1/3) and two crosses there, how many counters without crosses do we need to make both the same?" "*We need four because there are two without crosses and one with a cross. That's equal.*"

But for 2/5 and 6/13 (question 10), he believes that "*it's the same. There is as much risk on one side as on the other. Here there are six with crosses and seven without crosses, and there two and three* (thus a difference of one on each side)." "And if we compare three piles, 1/2 and 20/21 and 100/101, which is most likely?" "*There where there is one more out of one hundred, there is less chance of getting it than there is with the one more out of twenty* (after a long hesitation)."

We see clearly in these cases the reasons for the difficulties with questions of proportional probabilities. The two values of the favorable cases (A) and of the possible cases (B) are variable, and the child comes back to them or remains with them by a kind of oscillation or simply by false generalization of the same type that succeeded for him in the problems with a single variable. To take into account the parts and the whole at the same time, he begins by reasoning alternatively on the favorable cases (A) as if the whole (B) remained constant; then, on the unfavorable cases (A′) as if (A) remained constant. After this, if the method seemed to be inadequate to him, he made a limited comparison of the two parts, either (A — A′) or (A′ — A). We observe here an interesting progression from the easiest problems (1/2 equals 2/4) to the most difficult ones (disproportions 1/3 and 2/5, and the like, and, finally, 2/3 and 6/7). It is for this reason that Ram goes back and forth even in his comparison of 1/2 and 2/4 when considering the single unfavorable case in 1/2 and the two favorable cases with 2/4; while at the other extreme, Bis sees immediately the equality of 1/2 and 2/4 and 1/3 and 2/6, but for 2/5 and 6/13 he reasons about the difference between A and A′ which is one in both cases.

On the other hand, if we put the larger group into subgroups of the same size as the smaller one (for example, two groups of 1/3 in place of 2/6) usually the subject sees immediately the equality of the relationships, and it is by this means of correspondence that he finds for himself the correct solution in the early cases (cf. Kel). But in a curious way all we have to do next is to mix together the earlier divided groups for the child to miss the relationships that he just discovered, as if these discoveries were no longer valid (see, for example, Kel and Mey). This situation is an exact reproduction, with one notable exception, of what we observed with children under seven years in the experiments with simple correspondence: The child of five to six years is able to see a one-to-one correspondence between two equivalent sets, but no longer believes that they are equivalent sets as soon as the elements are not all in his sight.[1] However, at this early level, no logical-arithmetical operation is yet possible, while the children of nine to eleven years, whom we have just cited, are in full possession of logical and arithmetical operations. Clearly, therefore, it is only the fact of having to reason about two groups at once and compare their internal relationships that produces this kind of return to an intuitive state. In the case of double relationship, proportions require formal operations, and before these are possible, we get only intuitive solutions, quite comparable to those which preceded concrete operations.

We now clearly see analogies between these difficulties and those which showed the slow comprehension of the law of large numbers. This is due, on the one hand, to the fact that this understanding also assumes the use of proportionality. This is, on the other hand, a reciprocal part of the fact that the relativity of a same absolute difference (with Bis, for example, who had one unfavorable case in two, in twenty, and in one hundred) is considerably more difficult to understand with large groups than with small ones.

6. *The third stage: Solution of questions with two variables.*

In the third stage, finally, all of the questions asked yielded a general answer quickly, without the gropings seen in stage II B for questions of proportionality.

EN (10; 2): Question 8 (inequality of favorable), 1/3 and 2/3: *"This one* (2/3) *because there are two crosses."*

Question 9 (equality of favorable), 1/3 and 1/4: *"Here* (1/3) *because there are only three counters."* "But is there one cross in

[1] See Jean Piaget, *The Child's Conception of Number* (New York: Norton, 1965), chap. III.

each pile?" *"But that makes no difference; it's easier because here* (1/4) *there are four counters and three without crosses and here* (1/3) *only two without crosses."*

Question 10 (both unequal), 1/3 and 2/5: (after a short reflection) *"There's a better chance here* (2/5)." "And this way (1/2 and 2/5)?" *"Here* (1/2)." "Then why (2/5) rather than (1/3)?" *"Because there are two without crosses in each group, and there are two crosses here* (2/5)."

Question 7 (proportionality), 1/2 and 2/4: *"Here* (1/2). *No, it's the same in both."* "And this (1/2 and 3/6)?" *"The same."* "Why?" *"There are the same number with crosses and without crosses in both groups."*

DUR (10; 3): Question 7 (proportionality), 1/2 and 2/4: *"It's the same because there are four in one pile with two crosses, and two in the other with one cross."* "And that means?" *"In each pile there are twice as many."* With 1/3 and 2/3: "How many counters without crosses do we have to add so that the chances are the same?" *"Three* (thus 2/6) *because here* (2/3) *it's easier than here* (1/3)."

Bos (11; 0): Question 7 (proportionality), 1/2 and 2/4: *"It's the same thing."* "And this way (1/3 and 2/6)?" *"The same too."* "If I put 3/6 on this side and six with crosses on the other side, how many without crosses de we have to add to get the same chance?" *"Six."*

Question 10 (both unequal), 1/3 and 2/5: *"This one* (2/5) *is easier because there is one less than in the other* (2/5)."

LUT (12; 5): Question 10 (both unequal), 1/3 and 2/5: *"It's easier here* (2/5) *because it gives you two chances against three, and there, one against two. We would have to have four without crosses there* (thus 2/6) *for it to be equal. Then since there is one less here, it's easier."* "What about with 2/5 and 3/8?" (He counts on his fingers.) *"It's the same, no."* (Without any suggestion, he arranges the two sets in a form which allows him to make the correct decision.) *"No, two out of five is easier."*

"And now (2/5 and 4/9)?" (He leaves 2/5 arranged as we have just seen, and then he puts each piece with a cross from the set 4/9 opposite a piece without a cross, leaving the last one with two without crosses.) *"The chances are now equal since there is one without a cross on each side. Oh, no.* (He makes some calculations under his breath, but doesn't get out of the difficulty.) *Now I see it! It's better here* (2/5)."

"And what about with 1/3 and 8/17?" *"It's easier with 8/17. If it were the same, it would be 8/24."* "Where is there the better chance?" *"With 8/17 because there are double the number without crosses here* (1/3), *and not on that side* (8/17)."

"And if we put 2/3 here and a cross over here, how many pieces without crosses would you need to make the chances the same?" *"We can't. We have to add at least one, and then there are too many chances* (i.e., 1/2 is greater than 2/3, an incorrect response)."

Because of some difficulties in working with fractions, the subject determines the double relationships by a system of correspondences (Lut) when the proportions or disproportions are not immediately apparent. But contrary to the reactions of the subjects of level II B, the double relationship which they discover in this way is kept, even when the pieces are mixed. In this way they achieve, by putting together multiplication relationships, the linking of the parts with the whole, the source of disjunction, and the quantification of probabilities which the disjunctive mechanism allows.

In a general way, the study of random drawings (which we are finishing here) confirms what we saw about physical chance while still permitting us to show more clearly the operative mechanisms which are in action in the development of the notions of chance and probability. Since this was already the case in questions of physical probability (for which the factors are considerably more complex than in the questions related to the simple workings of chance), we note, therefore, that the fundamental probabilistic notions are developed only at the formal level, and we begin to suspect why this is so.

The essential reason for this is that the formal operations are psychologically operations of the second power, or operations which require previously learned operations, that is, concrete operations.

Since formal operations are of this nature, they are at first more abstract than the operations on which they depend, and attain, therefore, a hypothetical-deductive power unknown at the level of the concrete operations. This is why they must be acquired before arriving at the modality of the pure possible, a necessary condition for the development of the probable and the different connections (disjunction, etc.) indispensable for the logical development and the quantification of probability.

Since formal operations also consist of operations deriving from other operations, they furnish, on the one hand, the indispensable psychological instrument for the construction of certain operative mechanisms which are inaccessible at the time of the elementary concrete operations and which intervene necessarily in the development of probabilistic notions. It is in this way (as we have several times shown) that the law of large numbers assumes the use of proportions, and that these proportions themselves, which are the relationships of relationships (*des rapports de rapports*) require the intervention in the mind of the child of formal operations—no matter how concrete, from a logical point of view, the proportions remain. But above all, the achievement of probabilistic notions implies the ability to use combinatoric operations—combinations, permutations, and arrangements. We are going to see now that these operations constitute operations of the second order, and in their turn call upon formal mechanisms, psychologically speaking.

PART THREE

Combinatoric Operations

VII. The Development of Operations of Combination

Each of the preceding studies led us to expect the same conclusion: that the formation of the ideas of chance and probability depend in a very strict manner on the evolution of combinatoric operations themselves. It is from the ability to conceive mixture and interferences according to an operative scheme of permutations and combinations that the child comes to what is properly called the notion of random mixture, that is, to the notion of chance. On the other hand, the child constructs his notions of probability by his ability to subordinate the disjunctions effected within mixed sets to all the possible combinations, using a multiplicative and not simply an additive mode.

This connection between probabilistic notions and combinatoric operations is logically apparent. Thus the progressive mixing of marbles which we had the children work with in Chapter I is reducible to a series of operations of permutations. The only difference between this fortuitous mixture and the corresponding operations which followed is that the permutations which were done in a concrete manner in the mixture were not followed up in a systematic order, and especially since the marbles showed only one part of the possible permutations. The second restriction is more important than the first because, even without order, a complete realization of all the possible permutations would be equivalent to complete deductibility. In particular, we would be assured of seeing all the marbles return, sooner or later, to their starting places, and the mixture could then no longer be irreversible. But only a part of the possible permutations are achieved, and that is why there is chance, or irreversible random mixture. Even the understanding of the idea of chance implies, therefore, the understanding of combinatoric operations of which the fortuitous facts constitute a fraction of the accomplishment. On the other hand, and because of the fact that only a part of the possible operations interact, a second notion completes the notion of chance as soon as this combinatoric mechanism is understood: this is the idea of probability, that is, the relationship between the operations observed and all the possible operations. It is in this way that the mixture is oriented in the direction of the most favorable permutations, this probability consisting precisely in the relationship between certain partial permutations and the whole set of possibilities. In the same way, in the random drawings, the hand put into the sack finds only certain pieces among all those possible. In this last case the judgment of probabilities will consist in separating all of the possible combinations (additive disjunction, that is piece by piece, or multiplicative, that is by association of two, three, or *n* elements) and in determining the relationships between the observed combinations and the whole set of possible ones.

But if the logical relationship between probabilistic notions and combinatoric operations is apparent, it is no less necessary to find out if in the real psychological genesis (the actual foundation of operative relationships) the construction of ideas of chance and probabilistic intuitions actually needs the development of the operations of combinations. Above all, it is a question of understanding how the liaison between these two sorts of realities is achieved since, in appearance, they originate from such different psychological dispositions.

With this in mind, we have proceeded in the following manner. After questioning certain groups of subjects on questions of chance such as random mixture (Chapter I) or random drawings (Chapter V), we next asked them, without indicating any connection with the preceding trials, to permute or combine in all sorts of ways small sets of counters, and we then analyzed the different methods the children followed in these combinatoric operations. By then establishing the successive stages characterized by these methods, we sought to derive the relationship between these stages and the ones we found in the formation of probabilistic notions. The correlation was found to be very clear. Situations in which random mixture is present are structured according to the idea of chance after spontaneous combinatoric operations are developed.

In this chapter we will describe the formation of a first system of combinatoric operations, that is, the combinations themselves. In Chapter V we studied only the case of combinations of n things two at a time $\left({}_nC_2 = \frac{n(n-1)}{2}\right)$. We remember that the proof analyzed in Chapter V consisted in drawing pairs of pieces from a sack which had counters of different colors and predicting the colors of the draw as a function of the model distribution. On the other hand, for combinatoric operations, the following technique was adopted, one in which we will immediately notice analogies with the preceding ones. Piles of counters are put on the table (a pile of white ones, a pile of red ones, etc.) and the child is asked to construct as many pairs as possible, not repeating himself with already chosen colors (in certain cases we did admit pairs of the same color: blue with blue, etc.; other times we did not).

Following are the results obtained: In the course of stage I (up to about seven years as median), the child comes only to an empirical discovery of the combinations, without system and simply by means of groping (in the same way that he failed to have any systematic prediction of the composition of pairs or even in the random drawing of simple elements in the trials of Chapter V). In the course of stage II (from seven to eleven years), there is the search for a system, in the same way that the determination of probabilities marks for this level (Chapter V) the beginning of systematic quantification. It is interesting to compare the reaction to combinations during this stage II with the ability acquired (from about six or seven years) to construct simple correspondences. Clearly, the operations of combinations are later in development than these last abilities, and it is only in stage III (formal operations, from the age of eleven to twelve) that the child arrives at methodical and complete combinations, after some practice, during the apprenticeship which led to correspondences (from six to seven years of age).

1. *The first stage: Empirical combinations.*

Proceeding still by simple gropings, the subjects starting at six years succeed generally in finding all the combinations for four or five colors. The younger subjects, on the contrary, ordinarily do not get to this complete, even though empirical, construction. Here are a few examples beginning with a subject who does succeed with help in making a complete discovery using only three colors:

LEV (4; 8): Red (R), white (W), and blue (B). "These are little boys. Show me how they can leave two at a time in all the possible combinations. See, like this. Try it." *"R and B."* "Fine, and then?" *"B and B; W and W, and R and R."* "More?" *"R and B."* "And next?" *"R and B."* "You already have that one." *"Two W."* "Watch. You have that here. More? . . . The white is only with the blue. What other arrangements are possible?" *"W with W."* "You already have it. Any other? . . . What if we put R with W?" *"Yes."* (All the pairs are lined up.) "Are there any other pairs, or do you have them all? . . . How can you know if you have them all?" (He counts them.)

ROS (5; 9): White (W), red (R), and green (G): He puts together W and W, G and G, and R and R. "Any more?" *"This* (G and G)." "You already have that. Put two different colors together like this (G and W)." (He makes G and W and then R and G.) "Is that all?" (He looks around.) *"This* (W and R)." "How do you know if you have finished?"

We add blue (B). Ros makes B and B, W and W, R and R, then W and R, B and R, then W and R (which we eliminate), then W and G. "Are there any others?" *"No, I put them all out, all the colors."* "What did you put blue with?" *"With R, W, and B."* "What color didn't you put it with?" *"I put it with all of them."* "With green?" *"Oh, no."* (He puts B with G.)

SCHA (6; 3): Three colors, white (W), red (R), and green (G). He puts W and W, R and R, G and G. "What else?" *"This* (W and W)." "Didn't you already do that?" *"Yes."* (He takes it off and puts R with W). "And now?" (R and G, W and G, and again R and G.) "Do you have that already?" (He takes away the second R and G.) "Is that all?" *"Yes, we have them all. We need another color."*

We add yellow (Y). He puts Y with G, Y with R, W with G, W with Y, W with W, R with R, G with G, and W with R. (Each time he compares the original piles with the pairs he has made.) "Are there more?" *"This* (G with R)." (He stops.) "Any other ways?" *"No."* "You're sure?" *"Yes."* "How do you make sure?" *"I look at them."* "How many colors is the green paired with?" *"With four."* "And white?" *"Same."* "And yellow?" *"Three."* "Then? . . . Does that mean anything?" *"No."* (He is unable to see.)

PER (7; 3): Blue (B), white (W), and red (R). He puts B and W, W and R. "And if we continue?" *"W and W, B and B, and R and R."* "What next?" *"There are no more. We will have to start again."* (He does not see B and R.)

"Now what if we add the green (G)?" (He puts R and B, G and B, B and B, G and G, R and R, and W and W.) "Any more?" *"B and W, and W and G."* "That's not all." *"Yes, that's all."* "How do you know?" *"Because I can't do any more."* "Try again." (He puts R and B, takes it away and puts R and G.) "What did you do to find that one?" *"I looked closely and saw that there wasn't red and green."* (We now add yellow and brown, making six colors in all.) "How many pairs can you make with all these?" *"Twelve."* "Why?" *"Because there are six."*

Now we see how these subjects proceed. As long as it is a question of putting two of the same color together (i.e., R and R, G and G, etc.) they experience no difficulty in following the order of the given colors and thus exhaust systematically the series of available colors. It is clear, however, that there is not at this point any combination in the proper sense of the term, but only simple duplication of each of the terms of a series. On the other hand, as soon as it is a question of pairing one color with another one, the subjects do not look for a system but simply pair by chance any two colors, and then afterward try to see if there are any other combinations possible.

The combinations using just three colors already give instructive data. Lev, for example, working with white, blue and red, finds only red-blue then blue-white, but he fails to discover alone the pair red-white. Per, at the age of seven years, three months, in the same way, forgets to make pairs of all three.

As for the combinations of pairs with four colors, not one of the subjects succeeded in making a systematic arrangement of a single color with the other three, then another color with the other three, and so on. When it does happen that they put one color with two or three others in sequence, as Ros does putting red with white and then blue, it is simply perseverance and not an intentional procedure, since in these cases some of the same pairs (here white-red) reappear without the child noticing that he has already done that pair. Even more obviously, none of the subjects succeeds in finding for himself the six pairs implied by the four colors. It is enough to say that their method remains empirical, that is, proceeds by haphazard successsive pairings and not by the use of a plan which, even if inadequate, would give the first combinations.

This first stage is, therefore, comparable to what we saw in the beginning of seriation, for example, in the case of arranging sticks of different lengths: [1] Instead of constructing a scale of the whole group, the children of about four to five years of age simply juxtaposed a set of pairs made up of a short piece and a long one, without bothering with comparisons between the pairs. However, when they acquire the idea of seriation (six and a half to seven and a half years of age), that is, when they have an operative idea of groupings, the operations of combinations using three or four elements remains for this age preoperative, corresponding to the level of the first beginnings of seriation.

2. *The second stage: Search for a system.*

From the age of eight, the subjects were given the following experiment: Four different colored piles of counters were put on the

[1] See Jean Piaget, *The Child's Conception of Number* (New York: Norton, 1965), chap. VI.

table, and as before we let the child proceed in his own way to make pairs. Then we asked him if he couldn't find a quicker and more accurate system of doing it without forgetting any of the possibilities. After another trial, we then give him six colors to work with, asking him to arrange pairs without forgetting any.

PIE (8; 3): Four colors (which we call A, B, C, and D). He puts AB, CD, stops a moment and continues with AC, AD, CD (which he removes in a moment), and BD: *"That's all."* "How can you be sure?" *"I look here and here* (at the pairs he has made and at the piles). *Oh, still this one, BC."* "How many yellows (A) do you have?" *"Three."* "And reds (B)? *"Three."* "And blues (C)?" *"Three."* "And rose (D, but this time we hide the piles)?" *"Four."*

Six colors: He puts them together empirically. "Try to find a way not to miss any." (He starts again: AB, BC, CD, DE, and EF.) "Do you have them all?" *"Oh, no* (he puts B with D)." "Try again." (Again he puts AB, BC, CD, DE, and EF, and then he continues empirically.)

FRE (8; 6) finds empirically the six combinations possible with four colors: *AB, AC, AD, BC, BD, CD.* He tries for more and ending up with the same pairs, he concludes, *"It's finished. We can't make any more."*

We start again adding colors E and F. He works as follows: AF, ED, CB, FD, AC, EF, AE, CE, BD, BE, AD. "Is that all?" *"I think so."* "How many times did you use yellow (E)?" *"Five."* "And red (D)?" *"Four."* (He compares the piles and adds CD.) "And green (F)?" *"Three."* (He adds BF, CF, AB.) "Fine. Can you now find a way to be sure that you have them all?" (We start again: He puts EF, CD, AB, AF, BC, DE, and AD, and then continues empirically.)

DAN (9; 6): Four colors: He places AB and CD, next BC and AD. "Is that all?" *"Yes, no* (he adds AC and BD)."

Five colors: "Try to find a way to be sure that you get them all." (He pairs AB, CD, and DE, and then works empirically and finds AD.) "How many reds do you have?" *"Three."* "Roses?" *"Two."* "And blues?" *"One."* "And yellows?" *"One."* "Is that the way it ought to be?" (He adds blue and rose, and little by little comes to make all the missing pairs.) "And now how many are there of each color?" (He counts.) *"Four."* "Why?" *"Because there are four piles.* (He counts.) *No, there are five piles.*" "Why then do you have four pairs for each color and five piles?" *"Don't know."*

"Try to find a method so as not to forget any." (He starts with AE, then AB, BC, DE, then AC, BD, CE, and finally AD.) *"I don't find any more."* "How many rose?" *"Four."* "And reds?" *"Four."* "And blues?" *"Four."* "And yellows?" *"Three, oh, yes."* (He adds BE.)

CHEV (10; 0): Four colors: He puts together AB, BC, CD, then AC, BD, and AD. "How did you do it?" *"One after the other first, and then by looking to see what was missed."* "Try to find a simpler system." (He does it again, as before.)

Six colors: He matches AB, FE, BC, ED, CD, then skips as before, gets mixed up and continues empirically. "What if you took the first color and paired it with all the rest?" *"Oh, yes."* (He makes AB, AC, AD, AE, AF, but continues by going backward: FE, FD, FC, FB, FA.) *"Oh, no. I already have that one."* (He removes it and continues by skipping around.)

GIN (11; 0): Four piles: He puts together CD, AC, AB, BC, AD. "How many of each color do you have?" *"Three rose* (C), *two green* (D), *three whites* (A), *and two blues* (B)." "Are there any other pairs to be made?" *"No."* "If there are only two blues, can't you find a third one?" (He looks and finds BD.) *"Oh, yes, that's it. That way there are three of each."* "You've been guessing. Isn't there any way to do it so that you don't forget anything?" *"Yes, we could pair up three blues, three whites, etc."* "Try it." (He puts together AB, BC, CD, AD, AC, BD. "Try to find a better system." (He makes AD, BC, AB, BD, CD, and forgets AC.)

Six piles: First he puts BE, then CD, then EF, then AB, BC, DE, and EF. "Do you have them all?" *"Oh, no."* "Try to start at one end." (He does it again: FE, FD, FC, FB, FA, but continues from the other end: AB, AC, AD, AE, AF, then takes AF away.) "Do you have them all?" *"No, still this* (CD), *and this* (DE)."

ROS (12; 0): Four piles: He puts AB, CD, BD, AC, AD, and BD. "Find a better way to do it." (He puts three rose counters in a row and puts three other colors with them, but does not continue in the same way with B, and finishes empirically.)

Six piles: He puts AB, AC, AD, AE, AF, then BC, BD, BE, BF, but then goes to FE, FD, FC and finishes empirically.

We see the interest of these different trials of systems whose common traits teach us why the child (and the prescientific mind in general as well) has so much trouble in conceiving of the idea of combination and, consequently (or even more so), of the idea of chance. In fact, the general characteristic is to start from the additive idea of juxtaposition and not from intersection or interference, that is, from multiplicative association, a notion which, in the last analysis, the child discovers rather than presupposes. His ability to presuppose this multiplicative association would be necessary to construct a complete operative system. In this regard we should try to order the different systems we have just seen, going from the most rudimentary to those which come closer and closer to the correct method.

The method used in stage I consisted of making successive pairs independent of each other. This lack of system, in reality, rests on an implicit postulate which is precisely the belief in the independence of the pairs, as if each existed alone, that is, were formed of elements seen as having no relationship with the other pairs. It is this simple juxtaposition of pairs which is found in the most elementary example of this stage (a system which appeared, however, already in the course of stage I): The terms ABCD give pairs AB and CD with the other pairs to be discovered empirically (see how Pie begins).

A second system shows progress in the idea of intersection of the pairs, while remaining subordinated to the idea of juxtaposition. This is the most frequent system at this level (the one which Pie, for example, uses after the first one). For six colors, it consists of starting with AB, BC, CD, DE, and EF, the other pairs to be made empirically. There are, then, if it can be said in such a way, crossed juxtapositions.

A third system is noticed for its manifest intention: It consists first in pairing the two extreme terms to mark the necessity for a complete association, and only then in proceeding by crossed juxtapositions. Dan, after starting with the first method, passes directly to method three: with five items, he begins with AE, then makes AB, BC, CD, and DE, and continues by skipping each time one item (AC, BD, and CE), and ends up empirically. Fre chooses an analogous method.

A fourth method, which answers the same need to embrace the

totality of associations, but without being freed of juxtaposition, consists of proceeding with symmetrical pairs (see Chev, for example): AB then FE, BC then ED, and finally CD, the rest being found empirically. This tendency toward symmetry is so strong with Chev (as with many of the others) that, after accepting the suggestion of order AB, AC, AD, AE, and AF, he continues with FE, FD, FC, and so on, up to FA which he rejects, noting that he had it already (see Gin at the end of the questioning).

Finally, a fifth method, characterized by incomplete intersections, leads us to the edge of stage III. Ros discovers spontaneously the system AB, AC, AD, AE, AF, then BC, BD, BE, BF; but he is afraid to continue taking this direction, and reverts to the symmetry of method four. Others start in a similar way, but end up empirically.

We see that the characteristic of this stage is certainly to seek a system of associations instead of being satisfied with isolated pairs. But they still fail to discover a complete system. The reason for this is that the child wavers between juxtaposition (AB, CD, or AB, BC, CD) and symmetry (A → and F →) because he is unable to synthesize them into a single method of directed intersections whereby each term would be associated successively with all the other ones (AB, AC, AD, . . . and BC, BD, . . . etc.). It is interesting to note that these two tendencies (e.g., to a simple linear succession and to symmetry) hinder the subject from grasping the notions of probabilities founded on combinations (see Chapter V, Section 2).

3. *The third stage: The discovery of a system.*

From the age of eleven or twelve (sometimes even from ten) some subjects discover a system such that no pairing is skipped. This is in contrast with the systems of stage II which remain incomplete and must be finished with empirical searchings. Here are some examples beginning with an intermediate stage between stages II and III:

Sto (10; 7), who with six colors pairs first AB, AC, AD, AE, and AF, and then FE, FD, FC, FB, and stops before pairing FA. He next assembles BC, BD, BE, and ED, EC, EB (which he removes), and finishes empirically. "Do you want to start again to see if you can do better?" (Then he puts together AB, AC, AD, AE, AF, then BC, BD, BE, BF, then CD, CE, CF, then DE, DF, and finally EF).

CAD (11; 2) immediately finds a system for associating A with all the rest, then B with the following colors, and C in the same way. "How many of each are there?" *"Five reds, five blues, five greens."* "Wait a minute (and we hide the next pile). How many yellows?" *"Five also."* "Why?" *"Because there are five piles, no six, but we put each one with all the others."*

LAU (12; 3), using four colors, seems to start as did those in stage II. "Can you find a way of doing all of them, forgetting none?" *"Yes."* (He puts together AB, CD, AC, BD, AD, and BC.) "Are you just guessing?" *"Not at all. I looked for four of each and saw that three were enough. There would be four if we put yellow with yellow, blue with blue, etc."*

Six colors: *"There will be six of each, and not five."* He then puts five yellows (A) in a vertical column and joins with each one the colors B, C, D, and F; then puts five blues (B) in a parallel column and joins with them C, D, E, and F, then takes away the fifth one: *"No, since I already made that with the yellow ones."* He makes a third column putting four counters (C), no longer five, and joins to them D, E, and F. Then he takes away the fourth C saying, *"Oh, it reduces by one each time!"* He then puts two D and joins to them E and E; and then E and next to it F: *"That's it."* "Could you do it without the columns?" (He looks.) *"Oh, yes, like this* (AB, AC, AD, AE, BC, BD, BE, CD, CE, CF, then DE, DF, and EF)."

Faced with these simple solutions which show a complete comprehension of the problem (notably concerning the question of how many of each set of elements: five A, five B, etc.), we must ask ourselves why these operations of combinations are solved in this way only at the formal level (hypothetical-deductive operations beginning at about eleven to twelve years) instead of appearing at the level of concrete operations (seven to eight years).

Qualitative seriation and correspondence are, in fact, operations acquired at about seven years of age. The combinations achieved by Cad and Lau constitute only further correspondences, not one of which when taken alone goes beyond the level of ranges available to six or seven year olds. This is precisely the reason that from stage II the subject seeks to develop a system whose possibility he suspects as soon as he is capable of concrete operations. Why then must we wait for the formal level for such a system to be discovered? The answer is apparently that these correspondences are not independent of one another in the case of combinatoric operations, but that they do constitute a unique system of such a sort that what is done first determines what follows. To construct seriation or correspondence, it is enough to repeat the same operation a certain number of times and to invert it, taking into account all the terms (for example, to put B into a simultaneous relationship with A and C, having the form B larger than A and B smaller than C). On the contrary, however, to construct a system of all the combinations of pairs possible with *n* terms, it is necessary to coordinate several different series or correspondences and to anticipate the scheme of their relationships before making a proper construction. It is for this reason that this kind of system assumes the interaction of formal operations. In fact, the formal operations consist in a general way of operations of the second power, that is, operations bearing on other operations which can include concrete types of operations. In the particular case of operations of combinations, these imply a series of successive correspondences, and this means we are dealing with operations of the second power, related psychologically to the formal type of operations.

VIII. Operations of Permutation

We know that the possible number of permutations for n elements is $n!$. This means that for 2, 3, 4, 5, and 6 elements, we already have values of 2, 6, 24, 120 and 720. Therefore, there could be no question of having the children discover the actual formula of these permutations as a mathematical expression. On the other hand, however, nothing should keep us from asking the subject to find a system which allows him to find all the possible permutations when working with a small number of objects. This question, then, is the equivalent of the discovery of the formative operations themselves, as distinct from the formulation: Knowing, in fact, that two elements A and B yield only two permutations, AB and BA, they will see that the addition of a third element C gives three times as many, or $2 \times 3 = 6$, because we can place C with each couple (AB and BA) in three different ways (that is, CAB, ACB, or ABC, and CBA, BCA and BAC). They will discover as well that the addition of a fourth element gives four times as many permutations, or $6 \times 4 = 24$, because we can put D with each of the six triplets in four different ways. In fact, therefore, they will discover the law $n! = n\,(n-1)\,(n-2)\;.\;.\;.\;3 \cdot 2 \cdot 1$; even if they do not arrive at the explicit expression in symbols, they will at least succeed in seeing its operative mechanism, which is all that matters to us from the analytical point of view of probabilistic intuitions.

The formation of this operative mechanism, in fact, is of great interest for the study of the development of the notion of chance. In the final analysis, a mixture is always a system of permutations, and we have seen in our description of the difficulties the children have in understanding the nature of a mixture, that their failure to understand these permutations was pronounced. Clearly, this is not to say that the child ought first to have derived the system of permutations from objects aligned on a table so that he could next become capable of finding the mechanisms of permutations in any mixture. It is quite possible that he begins by grasping spontaneously the observed permutations among the objects, effecting only later a discovery of intentional permutations. But, in both cases, the fortuitous permutations become intelligible for him only as a function of his understanding of the systematic permutations. In fact, the true permutations are produced in the course of a mixture and are understood as fortuitous only to the extent that they enter as particular cases into the entire set of possible permutations. This set is precisely made up of the systematic permutations which arise from the operative mechanism described here.

Here is a point of fundamental importance for the psychology of chance. Rather than being imposed directly by the observation of experimental facts, the idea of chance assumes a structure, and as we have seen in Chapter I, this is necessary from the beginning of the observation of such facts, even in their most elementary form. It is only by reference to such operative systems, either by contrast or by analogy, that the subject recognizes the existence of fortuitous phenomena, and then so as to develop them, constructs notions of chance and probability. In the particular case of random mixture, there could be no understanding of what is properly called mixing —that is, of disordered and incomplete permutations, except by comparison with ordered permutations—that is, intentional, systematic and complete ones whose study we are concerned with here and which constitute an operative system.

1. *Technique and general results.*

We give the subject two toy men and explain to him that they are walking side by side. They can thus be placed in two ways, AB and BA. We next ask the child to do the same thing with two counters of different colors, putting one under the other to form two pairs AB and BA. Then we let him have three different sets each of a different color, from which he can draw the necessary elements for permutations ABC, BAC, and forth, which we ask him to do by placing the necessary colors in a row. If he succeeds in finding the six possible permutations, we finally add a fourth color to see if we can have him find in the same way the twenty-four permutations that the elements ABCD can give.

The answers obtained for these questions can be divided into the three usual stages: up to seven to eight years, no system; from seven or eight to eleven or twelve years there is the discovery of partial methods; and, after twelve years, progressive discovery of the law. But, contrary to the operative mechanism for combinations, acquired at about twelve years, the acquisition of permutations is not achieved before fifteen years. Out of the twenty subjects from eleven to fifteen years, only six discovered it spontaneously.

The first stage, however, divides into two distinct levels, I A and I B. At level I A, the child experiences a certain difficulty in understanding that several permutations can be made with the same elements and asks for some more counters of different colors to be able to continue. He has some difficulty recognizing the permutations with the counters that he observed with the toy men, and after making one or two permutations with the counters, he does not succeed in generalizing his empirical discovery. At level I B, on the other hand, there is further empirical groping and the discovery of some regularities (the inversion of a pair of positions or the possibility of starting twice with the same color), but without the subject's deriving any coherent system from the isolated discoveries.

In the course of stage II, we find a broad spectrum of comprehension in the search for a system including the three elements, and level II B can be characterized by the anticipation of the same possibility for four elements, but with no system as such being discovered at this stage for this number of elements.

Finally, in stage III, the system is discovered little by little, but as we have already noted, with a certain lag when compared with combinations because of the more rapid increase of the number of transformations as a function of the number of elements. We can thus distinguish two levels in stage III as well—a substage III A, in the course of which some erroneous systems are adopted before the correct one is found, the correct one not being discovered except as a result of suggestive questions. In substage III B (beginning at fourteen or fifteen years) generalization is systematic.

2. *The first stage: Absence of a system.*

Contrary to the case of combinations which the subjects succeeded in grasping rather quickly using empirical gropings (although with no system), and where they were able to construct most of the combinations for three and four elements, the children of this first stage show some difficulty in grasping even the principle of permutations. Here are some examples of level I A:

ELI (4; 4), whose answers to questions on random mixture we saw in Chapter I, Section 2. "You see these two little men who are shaking hands (AB). Can they be placed in a different way to take a walk? Like this (BA)?" *"Yes."* "And if we add the green one (C), how can they be arranged to take a walk?" *"Like this* (BCA)." "Another way?" *"This* (CBA)." "Another way? . . . Can the little green one change places?" *"No."* "Can they walk like this (BAC)?" *"Yes."* "And another way? . . . Like this (ABC)?" *"Yes."* "And what else?"

CHIN (5; 6) gives ABC, BCA, then again ABC and BCA. "Aren't they placed in the same way?" *"No."* "Look at this (we show him ABC and ABC and remove the latter one)." "And this (BCA and BCA), aren't they the same?" *"No . . . oh, yes."* (We remove the second one.) "Now find some new ways to do it." (We put the counters back in their piles and start again.) Again he makes ABC and BCA and stops.

OL (5; 8) permutes AB into BA. "Is there any other way to put them?" *"We've finished."* "Fine, now here's a third color (C)." "*Like this* (he makes ABC and CAB). *I think that's all.*" "You think that no other color can be moved?" (He adds BCA and ABC.) "You already made that one (ABC)." (He then permutes A and B and gets BAC.) "Have we finished?" *"No, we can still use some red ones."* "No, just with the colors I've given you." *"I think that we've finished."*

TIS (6; 10) is able to find ABC and BCA with the three pieces (by shifting A). "Can you still change them around in another way?" "*Yes* (CAB, ACB, and ABC)." "Do you already have this one (ABC)?" *"No."* "Look again." (He removes it.) "Can you do more?" *"I've done all that can be done."* "You're sure? Try again." (He makes CBA and BCA.) *"No, not this one* (BCA); *I've already made it."* "Can you do any more?" (Long reflection.) "*Yes* (BCA)." "Have you already made that one, or not?" "*Oh, yes* (he replaces it by BCA [sic])." "And that one there, do you have it already?" *"Oh, yes."* "Do you think there are any other arrangements?" *"No, there are no more. I made them all."* "How are you sure that you made all there are?" *"I want to see again if there are any more."* (He changes the counters around and makes CBA, BCA, and CAB, three of the permutations he had already discovered.) "Where do you look?" *"I look everywhere."*

Here now are a few examples from level I B where the child finds by gropings the majority of the possible permutations as well as some regularities, but still without searching for a total system.

BER (5; 6), with three elements, first simply permutes A and B, getting ABC and BAC. "Fine. Now try to do it in all ways possible." (He finds ACB, CAB, ABC, and CAB.) "Name all the colors you have changed." "*Blue and red* (changing ACB to CAB). *Red and green* (CBA). *Green and blue* (BCA). *Blue and red* (BAC). *Red and green* (ABC). *Green and blue* (ACB)." "That's fine. How many times did you switch them?" *"Four times."* We note that in this explanation by Ber, each initial color corresponds to the permutations of the two following ones (CAB and CBA; BCA and BAC; then ABC and ACB), but the subject himself did not perceive this regularity.

GRAM (6; 7) has a great deal of trouble in taking the initiative. During the whole first part of the experiment, we have to push him to find the next move after each permutation of the three elements (up to six possible ones). "Fine, now try it alone." *"There* (ABC, CAB, and BCA)." "What do you do to make sure that you have them all?" *"I don't know. I look."* "And then?" (After long pauses, he adds BAC, ABC, and ACB.) "Any there that you have already made?" *"No, yes* (he takes away ACB)." "Can you make any more?" *"No."* "Could we find a way to be sure we have them all?" *"Yes."* "What is it? *"Don't know."*

GRAM (7; 3) finds empirically five of the six possible permutations for three elements. "Do you think there are any more?" *"I want to try."* (He starts over again and finds the same ones.) "Do you find any more?" (He reflects for a time then makes one he already has.) "Was that already there?" *"Oh, yes."* "Are there any more?" *"I'll try."*

We note here that not one of these subjects was capable of finding a system enabling him to get all the possible permutations for three elements. It is, therefore, useless to advance to the experiment with four elements because of the lack of an operative mechanism allowing us to study an eventual generalization. At level I A the child is not even capable of finding all the possible permutations by gropings. He will succeed at this in level I B at times, but still without an intentional system. Even when he gets to the end, having found the six possible permutations, he is not certain that he has finished. In certain cases (cf. Ber), by the action of assembling the counters, he succeeds in finding certain effective regularities, but this is not conscious and shows no systematic intent.

The first problem posed by such facts is, therefore, to understand why the children (level I A) have so much difficulty in finding all the permutations of three elements while the combination of distinct elements into pairs is so easy for them. There is a simple answer. A given series forms a perceptive or intuitive whole, relatively closed and rigid. To permute the elements of this series presupposes a certain mobility, either of an operative nature (stage II), or at least deriving from what we have called elsewhere decentration found in articulated intuitions (substage I B). In the absence of this mobility, the original order of the counters is a problem because it can seem to be the only possible order (see, for example, Chin who three times in a row remakes the same ordered triple), while combinations are formed by the addition of new elements and are, from this point of view, an easier activity. We see the proof of this in the tendency of the children (for example, Ol) to ask for new colors to allow them to construct new permutations because it is easier to add some terms to the ends of a given series than it is to change its order.

As for the absence of a system (which is still true of level I B), this is no doubt due to irreversibility which is a characteristic of the thought in all the stages preceding the appearance of the first logical operations (addition or correspondence), and which prolongs the rigidity we have just spoken of in relation to level I A. The ability to find empirically the different possible permutations consists in going ahead with the requested changes, while finding a system means always going back to the starting point as well as making the changes. A system, in fact, is the union of all the possible permutations in both directions (we know that the whole set of permutations constitutes what is called in mathematics a permutation group), and it is, therefore, natural that thought, which is irreversible concerning elementary operations, will be all the more so in considering a whole set of permutations that is formed by their union.

It is even more interesting to compare these reactions with those we described when considering random mixture (Chapter I, Section 2). We have here, in fact, the same subjects who were questioned about the progressive mixture of marbles in the tilting box and in the operations of permutations, and it is their two sets of reactions which have allowed us to develop the stages of Chapter I and those of Chapter VIII. We remember that from the point of view of random mixture, it is in the course of stage I that the child refuses to believe in a real mixture and imagines, therefore, that the tilting box ought to produce an immediate return of the marbles to their starting places. In the present experiment, on the contrary, the child shows himself to be incapable of the elementary reversibility which would consist either of permuting according to all the possibilities, or of retracing mentally all the permutations actually found so as to derive the beginnings of a system. But as we have seen in the case of the random mixture, the return to the starting positions that the child postulates is in no way a reversible operation. It is, on the contrary, only the expression of the rigidity of the initial order which is opposed to fortuitous permutations which constitute random mixture, an order to which the child returns precisely because he lacks the ideas of mobility and of reversibility. We now have a well-founded proof of this interpretation: To the extent that the subject does not succeed in making permutations or in finding a system corresponding to the permutations made during the experiment, he believes that the marbles will return after being mixed. The reason for this is precisely that he does not conceive of the random mixture as a set of fortuitous permutations, while it is to the extent that he will be able to conceive of the permutations as a system of reversible operations that he will understand the irreversibility of the incomplete and random permutations which are produced in the random mixture.

3. *The second stage: The empirical discovery of partial systems.*

The second stage presents a whole spectrum of reactions, ranging from the first elementary forms of recognition of regularities to the anticipation of organized systems which will characterize the third stage.

The most elementary reaction seems to be found in those subjects intermediary between the two stages who discover no systems by themselves, not even partial ones, but once having found the six permutations for three elements show an impression of regularity, a *Regelbewusstsein,* in the sense of K. Bühler, when we have them count the pieces according to the colors shown in each column of the picture:

WEB (6; 7), with three colors, makes ABC, CBA, BCA, and CAB, then (without seeing that he has already made them) makes BCA and CAB, and says, *"There, I've done them all."* "How are you sure? . . . Is that all?" *"No, I can still make some more* (BAC and CBA, this last one already made)." "That one (ACB), do you have it already?" *"Oh, no."* "And BAC (already there)?" *"No."* "Are there more?" *"No, that's all."* "How can you be sure?" *"I look at each line. There are no more; it's finished."* "But if you tried, would you find no more?" *"Maybe."* "I'll help you a little. Watch. How many times did you put the red one in the first line (we point out the first column of the six permutations)." *"Oh, I know, I put it here twice* (first column), *and that made six times, and the blue also six times, and yellow the same. Then there are three little men and six lines* (six permutations) *because I made them six times."* Web, therefore, certainly has an impression of regularity, but is not yet able to generalize with four elements (with this number he makes only a few permutations and those purely by gropings), nor is he able to understand why three elements give six permutations.[1]

Next come the subjects who belong properly to stage II:

BOR (8; 7) finds empirically the six permutations for the three elements and then starts again with some idea of order, the first elements chosen being A, B, C, and then C, A, B. In the same way with four elements, if he finds only eight permutations, he seems to have looked for the first elements according to a principle of repetition: A, B, A, B, C, D, C, D.

JOS (9; 2) discovers fairly rapidly the permutations for three elements. "Can you find any more?" "I'll try." (He starts again and finds the same ones, without any order.) "Now try it with four counters." He finds seven permutations and then says without prompting, *"There's only a single gray one there* (first vertical column) *so we can start again with a gray one."* Jos then finds three new permutations, but he does not try to see if the colors are now equally distributed in the four columns (he doesn't even look again in the first one).

[1] Cf. later the case of Sto (at the end of the questionings about three elements).

GEO (9; 6), with the three elements, finds the permutations ABC, CAB, and BCA by putting A at the head, the middle, and the end. "And what else?" *"There is no more to be done."* "Try again." He finds BAC, CBA, and ACB by doing the same thing with B (with an inversion of C and A in the fifth permutation to correspond with the inversion of C and A in the second one). "What are you looking at?" *"This* (he points to the vertical columns, but without trying to see if the colors are equally distributed)." "You could start over again to be sure you have done all the possible ones and not done any twice." He starts again with the same gropings after the third permutation, checking after each trial, but without counting the colors.

Next is a somewhat higher level in which the child starts in the same way but immediately finds an order when he begins to permute three elements, without, however, generalizing with four elements.

BAC (9; 10) finds ABC, BAC, and CBA, then gropes to finish. He starts over again following the order of the first elements: A, A, B, C, B, C, then finally finds the order A, A, C, C, B, B by permuting each time the second and third elements. But he does not generalize this method using four elements.

STO (8; 4): Two counters: he switches AB to BA. "And is there any other way of putting them?" *"No."* "Try with three." (He makes ABC, ACB, CAB, CBA, and BCA.) "Is there still another way?" *"No, there are no more."* "Certain?" *"Oh, yes* (he makes BAC)." "How many are there?" *"Seven, no, six."* "Can we make a seventh one?" *"No."* "Can we be sure in advance?" *"We can still make ACB; oh, no, it's already there.* (He does a few permutations over again.) *That one's already there; that one too, and that one."* "How many are there of each color in the first column?" *"Two blues, two yellows, two rose. We could add a third one since there are three colors.* (He looks around.) *No."* For three elements Sto finds on his own the system A, A, B, B, C, C, but without suspecting it at first, and not noticing it until afterward.

Four counters: He finds seven permutations by inverting the first two elements and then the two in the middle, then the outside ones, but without finding, without even looking for, any regularities.

GEN (10; 2) finds empirically the six possible permutations with three elements, and then starts over again by adopting at once the order A, A, B, B, C, C for the first elements. For four elements he proceeds empirically and stops at seven.

AP (10; 5), after a similar beginning, finds only three permutations for four elements. "Can you make more arrangements with three than with four?" *"It's easier with three than with four. You can make more arrangements with three."* "Why? . . . And how many with five?" *"With five you can make more because there are more piles. I was mistaken; there are more with four than with three."* Next he finds seven (with four elements) but with no order. In the course of long questioning, and with all sorts of suggestions, at the end we have gotten him to find the twenty-four possible permutations: "Did you simply look at the colors, or did you count them?" *"I looked at them."* "In each row (vertical), are there the same number of colors?" *"No, there are more red ones."* "Count them (first column)." *"Six, six, six, and six."* "And is it the same thing in the other columns?" *"Yes* (hesitates and then counts). *Yes, six also."* "Why?" *"Because in each row* (horizontal) *there is one counter of each color."*

Level II B begins next when the child begins to generalize spontaneously with four elements what he discovered with the permutations of three elements:

REN (10; 6) first believes that there will be more permutations with four elements than with three, but he finds only four permutations placing color A at the beginning, then at the end, then in the two middle slots. *"It's finished."* "How do you know?" *"I put green once first, once last, once second, and once third."* "With four counters, that would mean only four different ways." *"There aren't any more."* "Then are there more ways with four or with three counters?" *"Fewer with four."* "Why?" *"If I change it again, green will come back to the same place."* Later he discovers two more permutations. "Is that one new?" *"Yes, it hasn't been done before."* "Why?" *"When I start with yellow, I look to see if the other time I did the same one; I look at the second and then at the third and the fourth.*" "And how do you find a new arrangement?" *"When I use blue to start with, I watch carefully so that it doesn't have the same colors after it* (as in the preceding ones)." Little by little he finds thirteen different permutations. "Are there many more?" *"We can find some more."* "About how many?" *"Five."* "Why?" *"Just because."* "How many times did you start with green?" *"Four times."* "With yellow?" *"Two."* "With blue?" *"Five."* "And then what color will you continue with?" *"With yellow. If we can do it four times with green, we can surely do it four times with yellow."* "If you can get to four, will you be able to do more?" *"I couldn't do any more; I'd come up with the same arrangements."* "And the red ones?" *"I think I didn't make all the red ones with just one."* On the other hand, he is not worried about the fact that there are five with blue.

DEL (10; 5) finds with three elements: ABC, BCA, and CAB. "Can you find some more?" *"Yes* (ABC)." "That one is already there, but can you find some more by starting with A?" *"Yes* (ACB)." "Now find a way to place them so as not to leave out any." *"Oh* (ABC, ACB, BCA, BAC, CBA, and CAB)." Thus he puts in the first column: A, A, B, B, C, C.

Four elements: He finds four permutations starting with A, three starting with B, then finds a fifth one starting with A, and three new ones with B, then six with C (thus five, six, and six, but he does not complete his scheme by looking for a sixth permutation starting with A, and he does not bother with D at all).

LEI (10; 6), with three elements, finds three permutations beginning with A, with B, and with C, and then discovers the three remaining ones starting with the same terms but with inversion of the last two. "And then if we use four colors, they will make how many arrangements?" *"Eight."* "Go ahead." He begins by groping and then declares, *"We can make four green, four reds, four yellows, and four blues* (at the beginning of four distinct arrangements)." But he continues empirically and does not see until the end that there are six permutations for each starting color.

DUC (11; 6) first finds empirically and then systematically the six permutations for three elements. For four elements he finds eleven by using two-by-two permutations. "How many more are there?" *"Don't know. At least nine."* He starts again putting A in the first column six times in a row, then permuting by gropings. "Is there a trick for doing it better?" *"Put the second color in the same way. I hadn't thought of that."*

REST (12; 5) for four elements tries to put the same color in diagonal: in the first, then in the second, third, and fourth rows; he then does the same for the second color, etc.

HERT (12; 6): *"There must be more arrangements with four, I believe"* (he begins empirically)." "Might there be a trick to find them better?" *"Maybe we could put together all those that start with the same color."* He starts to do this and then says, *"There are more arrangements starting with red. I want to do the red ones."* He makes six permutations beginning with red and then starts with green. "How many times could we start a group with green?" *"Same as with red. It must be the same thing."* "Could we build them in a better way?" *"Yes, we could start with A* (red), *then put B* (green), *then underneath another B, then we could change places."*

ALM (12; 5) empirically finds nineteen permutations with four elements. "Do you think there could be a way of doing it better?" *"Yes, put those beginning with the same color together* (he counts five A, four B). *I ought to be able to make another of those* (he finds two more, giving him six B)." He does them all over again, starting six times with each color, but he does not order the last two colors.

JEN (12; 6): Same reactions, but he only finds five red ones, and says, *"It's not very possible because there ought to be the same number of colors."* "Maybe seven?" *"No, because there is an even number of counters* (four) *and an even number of ways of arranging them."*

POUG (13; 5): Same reactions: *"We've got to have the same number for all of them."*

We have been careful to cite all these facts because for us they have a certain interest, as much for the psychology of the operations in general as for the notion of chance. They allow us to observe, in fact, a very slow and continuous development of the consciousness of order and regularity characteristic of these operations from the moment when that consciousness consists only of a vague and intuitive anticipation (Web) until the time when it is reflective and deductive (Poug).

This leads us to a first remark concerning the relationships between these facts and those related to random mixture as conceived by the same subjects. As long as the child constructs his permutations empirically, as in stage I, without being conscious of any rule (in other words, proceeding according to the chance of his own gropings), he attributes to the random mixture (that is, to the number of fortuitous permutations), a relatively ordered structure contrary to the idea of chance. On the other hand, at the moment that he suspects and to the extent that he discovers the operative regularities proper to intentional permutations, it is then that he understands the fortuitous or uncertain nature of permutations which are inherent in random mixture. But as we have already noted at the end of the preceding section, this double correlation is quite natural. When the child, during the first stage, constructs his permutations by fortuitous gropings, in reality he is seeking to guess at an order, and he believes he has reached true order even though this order remains, in fact, subjective. In the same way, when faced with a random mixture, he seeks to conserve an order and foresees a return of the pieces to their initial order. However, to the extent that the progress of his operative intelligence allows him to anticipate and to discover in part a true order in the permutations that he constructs intentionally, he understands by this very action the absence of order in the fortuitous permutations which constitute random mixture. Moreover, in the same way that the progress of this operative regularity is very slow in its understanding of permutations, the progress of understanding chance in random mixture is also quite laborious, since the child remains for a long time attached, in spite of himself, to the idea of an underlying order (subjective, but projected on the actual) that he attributes to the trajectories of the marbles as they are mixed.

In the light of all this, what are the mechanisms at play in this progressive consciousness of regularity and how are we to explain in detail their correlation with the notions related to random mixture or chance? In the intermediate subject Web it is simply a question of an observation after the fact provoked by the experimenter but which is accompanied by an intuitive anticipation of the need to have the same number of all the colors. With the subjects Bor, Jos, and Geo, on the other hand, there is a beginning of regularity in the disposition of the first elements of each permutation of the three elements or in the order of the permutations (for Geo), showing even a beginning of transfer for the four elements (but without intentional generalization). These beginnings of order lead Jos to remark that if a single group begins with a certain color (for example, gray), there must be others of the same. Therefore, there is in this case a beginning of symmetry in the divisions. With the subjects Bac, Sto, Gen, and Ap, the order A, A, B, B, C, C, is discovered for the series of initial elements of each permutation of the three elements when, after constructing the permutations empirically, the child starts making them over again, but this time seeking a system. He does not, however, always apply this system to the case of four elements with an intentional generalization. When working with permutations of three elements, it happens that he discovers, in particular, that there will be the same number for each color in the vertical distribution (cf. Duc), but he does not apply this discovery to the groups with four elements; Ap, for example, believes that *"there are more red ones"* before counting. At level II B, finally, the discoveries made for three elements are applied to the permutations for four elements. In this regard we notice two correlative processes: first, a progressive consciousness of symmetries or equalities of distributions: *"If we can do it four times with the green, we certainly must be able to do it four times with the yellow ones* (Ren)"; with the green *"as with the red, it's got to be the same thing* (Hert)"; *"five A, four B. I must be able to make another one* (Alm)"; *"it's not very possible* (five and six) *because you've got to have the same number with each color* (Jen)"; and, finally, *"it has to be always the same with each color* (Poug)."

Secondly, now we have the beginning of a discovery of a system: first the equalization of the number of permutations beginning with the same color (Ren, Del, Hert, and Lei); next the repetition of the color of the second element (Duc and Hert or the using of a diagonal by Rest). It is only in stage III that the system sketched here will be generalized.

In viewing the correlations with the development of chance, this progressive discovery of regularities is extremely significant. On the one hand, we have just seen that this is opposed to operative regularities, since the permutations which interact in random mixture are considered fortuitous because they do not follow each other with any system, but constitute only isolated modifications not related to each other. On the other hand, it is to the extent that these operative regularities are understood that the fortuitous permutations taken as a whole are seen by the subjects as having to end up with symmetries and regular divisions. Now we remember that, just as in the bell-curve distribution, the symmetries are acquired only gradually, in the regular distributions (rotation of the disc, random mixture, etc.) the distribution of the whole is only perceived later and very gradually. We now understand the reason why: For a distribution as a whole to be conceived of as symmetrical or regular, it must be assimilated into an operative scheme which is the only way to explain the play of compensations leading to this symmetry. In other words, the chance of singular, fortuitous permutations is understood only by contrast with or by opposition to systematic permutations; but the probability of the whole is grasped only by analogy with the system of these operative permutations. To be noted, in particular, are the judgments of probability which derive from the developing operative regularity: Ren declares that he is going to start the next permutation with yellow because he has two yellows and four greens. He is, therefore, sure to find new permutations starting with yellow and green when there is a large number of drawings.

In short, the development of the operations of permutations explains the development of notions related to random mixture, and it is because these operations are so late in appearing that the interpretation of random mixture for the subjects is so difficult. But we must ask ourselves now why these operations of permutations are themselves acquired with such difficulty. A permutation is a change of order. And we have already seen elsewhere that order is relative to operations of placement, and the change of order is relative to inverse operations of displacement (that is, movements conceived independently of measure).[1] On the other

[1] Jean Piaget and Bärbel Inhelder, *The Child's Conception of Space* (New York: Norton, 1967).

hand, in studying the notion of movement, the moving of an object into successive positions such as A→B→C→D, and then D→C→B, then B→C, and finally C→B→A, and asking if in these cases there is the same distance covered in both directions as the counters are moved, we observed the same difficulties as if four objects (A, B, C, and D) were changing positions and finally coming back to their original places (hence an equality of the sum of the distances in one direction and the sum of those in the opposite direction).[1] We understand then immediately the relationship existing between these operations of permutation, the representation of the trajectories in a random mixture, and the composition of these displacements made in two directions. The general difficulty consists in uniting into a single system of thought several different systems of movements, either successive but having to be compared with each other as if they had been simultaneous, or simultaneous and having to be interpreted as if they were successive. From this comes the tendency of representing to oneself the tragectories in the mixture (Chaper I) as all being oriented in the same direction (and not in both directions simultaneously), and the difficulty of visualizing all the possible permutations of three or four elements—that is, a system of displacements which are simultaneously oriented, some in one direction and others in another.

4. *The third: The discovery of a system.*

The problem of permutations is somewhat more complex than that of combinations. Thus it is not resolved beginning with stage III—that is, at about eleven to twelve years—but only at level III B, which is at about fourteen to fifteen years. In the course of substage III A, however, we do find cases of more and more systematic solutions, a few examples of which follow:

LAY (12; 4) is moving around three counters A, B, C (with no other counters present so as to avoid empirical gropings). "Can

[1] An experiment successful only with subjects of about eleven to twelve years. Jean Piaget, *The Child's Conception of Movement and Speed* (New York: Basic Books, 1969).

we predict how many different ways we can arrange them?" *"With three counters there ought to be six different ways because each one can change places twice."* "How did you find that?" *"I arranged them in my head."* "Do it on paper." *"ABC, ACB, BAC, BCA, CAB, CBA."* After the third one, he wanted to make CAB, but suddenly he said, *"No, I've got to do them in order. We can put each color twice in a row."* "For four colors, can we tell ahead of time?" *"For three there were six; for four, there will be maybe eight."* "For certain more?" *"Oh, yes; with more elements, there are more possible arrangements."* (He puts ABCD, ACBD, ADBC, ADCB, ACBD and ABDC.) *"See, there will be twenty-four."* "Why do you say that?" *"It's simple. I started with the same color* (A). *That made six. Therefore twenty-four in all because for each color, we get six, and six times four is twenty-four."*

REM (12; 4) finds empirically the six permutations for three elements. "Are there any more?" *"First I have to see what I've done."* (He looks at his groups in which we see no system.) *"No."* "And now with four colors." He does the first four empirically. "Can we know how many there will be?" *"We would have to calculate."* (He remakes four of the permutations, starting with A, then with B.) "How many times did you start with red (A)?" *"Four times."* "And with blue (B)?" *"Twice."* "What if you started always with the same color?" (He finds six different permutations starting with A.) *"It will be twenty-four because four times six will give twenty-four."* "And with five colors?" *"Maybe eight per color, giving about forty."* "You can try it." (He begins systematically starting with A and switching B, C, D, and E.) *"Oh, I've found twenty-four with four colors, so I can make five times twenty-four, one hundred and twenty."* "And for six colors?" *"Six times one hundred and twenty."* "And for seven?" *"Seven times six times one hundred and twenty."* "Can you make a rule now?" *"We always multiply by the number that comes next."*

HUT (13; 6): For three, *"it makes six arrangements."* "How did you do it?" *"I switched them around in my head."* "For four colors?" *"I think a dozen."* "Why?" *"Because there is one color more which can change places with all the rest."* "Try it." He gropes and then says, *"No, we have to follow an order and that makes six times with each color, thus twenty-four in all."*

Here now are examples from level III B, beginning with a case which is intermediate between the two substages:

BIL (15; 0) begins by finding empirically the first permutations for three elements and then proceeds with an order. *"That's all. We can make six of them which surprises me because I was expecting to find nine."* "What about with four elements?" *"I believe there are sixteen, that is four possibilities for each one; I can change their places four times.* (He makes four permutations beginning with A.) *That must be all. We'll see.* (He find two more.) *O.K., there are twenty-four possibilities."* "What about with five elements?" *"For three pieces, six arrangements, for four, there are twenty-four. With four A's in the first position, I will get six possibilities. If I put them in the fifth position, I would have then six more possibilities. Next we have to add: it will be about fifty possibilities.* (He makes the trial and starts with four A's.) *I didn't do it right* (he adds two). *Perhaps it will be six times six, thirty-six. No, it makes exactly one hundred and twenty possibilities."* "Why?" *"For the six A's, I have twenty-four possibilities, thus five times twenty-four is one hundred and twenty."* "And with six elements?" *"Six times one hundred and twenty is seven hundred and twenty."* "And with seven elements?" *"Seven times seven hundred and twenty. For eight, we multiply it by what we got with seven."*

WEI (13; 9): "With three?" (He makes ABC and ACB.) *"With one color* (A), *I can make two. Using three will give six."* "And with four?" *"All I have to do is to add each time one color to the front. That makes four times six, twenty-four."* "And with five colors?" (He puts a fifth color in front of each of the permutations done with four elements, and concludes, five times twenty-four.) "And for six colors?" (He constructs a new scheme, putting in front of a column of six F's five columns of E's, and so forth, and says, six times five times four times three times two.)

BRES (14; 0) makes the first four permutations for three elements. *"Oh, I see already that I can put each one twice in the first position. That will make six."* "What about for four elements?" *"I can put the new color first and I can put it there six times. Then I can use the same system for each color. That will be twenty-four possibilities."* "For five elements?" *"There are one hundred and twenty ways.*" "Why?" "*If there are two colors, then we have one times two. If there are three, then one times two times three. If there are four, then one times two times three times four. If there are five, then one times two times three times four times five. And we can continue in the same way.*" And this is a student who did not know the law before the questioning.

We note that the reactions of level III A constitute a simple generalization of the partial systems already developed in the course of stage II: After establishing for n elements that with an initial element (A) there corresponds $(n - 1)!$ possible permutations of the other elements (for example $n - 1 = 2$ for the three elements A, B, C, when A is in the initial position, yielding ABC and ACB), the subject multiplies these $(n - 1)!$ permutations by the number of elements able to be placed in front of them (hence, for three elements three times two equals six), since he has known since stage II that for each beginning element there will be the same number of possible permutations. The discovery indicative of level III B (and which appears in the course of substage III A) is, therefore, that for four, five, six, . . . elements, the total number of possible permutations is simply determined by the multiplications of four times five times six times . . . , starting from the two times three permutations for three elements. This discovery results, as does the preceding one, in a reversal of directions when compared with empirical gropings: Instead of going ahead with the successive permutations, it is rather a question of preceding the permutations of n terms by the number of elements able to be put in front, and in this way multiplying these numbers instead of being content with accumulating them. But at this level III B the reversal of direction already sketched out earlier becomes systematic and allows the generalization one times two times three times . . . , that is, the discovery of the general law of permutations (see Wei and Bres).

Independent of the connections with the interpretation of random mixture and chance, the development of these operations of permutations poses an interesting problem for the psychology of operations in general. As mathematicians have shown, permutations form an elementary operative group. We can compose two permutations into a third, invert a permutation, associate successive permutations in different ways, and keep the identity of nonpermutation. The constitutive operations of this group consist in changes of order, that is, in operations which seem to be easily understood and easily done. We note, on the other hand, that the number of possible permutations is discovered by the child of this third stage by a simple trick of successive multiplication (two times three times four times . . .). Multiplication itself is a simple operation consisting essentially of correspondences. Why then are permutations acquired only at a formal level (stage III), while operations of order and the multiplicative correspondences are established from the level of concrete operations (seven to eight years)?

The answer to this is simple and confirms everything we have seen up to now on formal thought. If a change of order is in itself an elementary operation (or concrete), a multiplication of changes of order is, on the other hand, no longer a simple operation, since it is then a question of an operation bearing on other operations, that is, an operation of the second power. Formal thought is precisely characterized by operations of the second power, or operations which organize into a superior whole elementary operations which serve as their content. This, then, is the reason for the late character of the operations of permutation, as with general combinatoric operations. It is true that we can thus conceive of formal operations as built by the logic of propositions, in opposition with the logic of classes and relations which derive from the content and not from the pure form of the propositions. In this case, formal thought would have as its criterion the implication (if p then q), as opposed to operations of sets (inclusion, etc.), and relations (order, etc.), and number (multiplication, etc.). Permutations would seem, therefore, to call on these latter operations and no longer on the purely propositional. But in reality there are two forms of implications: the implication between propositions which characterizes formal thought in all cases, and the implication between operations which constitutes either a particular case reducible to the preceding one (even though these properties of the individual case are expressed by propositions, and even when each proposition of the general case expresses in its content an operation on sets or on relations, and each formal link between propositions is seen as an operation of the second power in so far as it is an interpropositional operation). From this point of view, the discovery that the operations of permutations consist of multiplying, according to certain mathematical relationships, a certain number of changes of order means, therefore, that we are dealing with an operation of the second power. At the same time we are dealing with the establishment of implications between operations, all of this corresponding doubly with the criteria of formal thought and explaining the late development of the system of permutations.

As for knowing why the discovery of permutations is later than that for combinations, satisfying both these criteria, the reason for it no doubt is that combinations consist simply in associations effected according to all the possibilities, while permutations, which are much more numerous, imply an ability to relate according to a mobile system of reference (transformation of the starting order for variable initial elements). While the operations of combinations constitute a simple generalization of those of multiplication, permutations furnish the true prototype of relations of relations, or operations bearing on other operations. We can then consider level III A as marking only the beginnings of formal thought, and the substage III B as the level of equilibrium of this mode of structuring.

IX. Operations of Arrangement[1]

After analyzing combinations and permutations, it may be interesting to examine the operation which is the synthesis of these two and which is generally called arrangement.[2] In the arrangements of two terms A and B, taken in pairs AA, AB, BA, BB, there are, in fact, simultaneously combinations (AA, BB, AB) and permutations (AB and BA). But the study of these arrangements will not simply be a means of allowing us to organize what we have just concluded in our research above; it will also lead us to a correlation which seems to exist between the combinatoric operations and the progress of the idea of chance. In this regard it will be sufficient to study the intentional arrangements made by the child along with his ideas on the resulting fortuitous arrangements, for example, from a random mixture of cards drawn in pairs. Now the exigencies of the analysis led us earlier to dissociate all the

[1] With the collaboration of Marianne Denis-Prinzhorn and M. Laurent.

[2] By "arrangement" as opposed to "permutation" or "combination" the authors mean that repetition is allowed. For example, if two-digit numbers are to be constructed from the three digits, 1, 2, and 3, there are nine possibilities 11, 12, 13, 21, 22, 23, 31, 32, and 33. With permutations of the three digits 1, 2 and 3, taken two at a time, there are only six possibilities since 11, 22 and 33 are not permissible (translators' note).

following: the study of operations of combinations (Chapter VII), the study of fortuitous combinations (Chapters V and VI), the study of the operations of permutations (Chapter VIII), and those of fortuitous permutations (Chapter I). In the light of these analyses, we will be able to describe simultaneously the reactions of children vis-à-vis operations of arrangements and fortuitous arrangements.

1. *Technique and general results.*

The material used is a deck of seventy-eight cards: twenty-six of them have the number one; twenty-six the number two, and the last twenty-six, the number three. For the children who do not yet know the different numbers that can be made with two of the three numbers, we use a second deck of cards whose three series are twenty-six locomotives, twenty-six railroad passenger cars, and twenty-six freight cars. Finally, a machine to deal cards allows us to take two cards at a time from one or the other of the well-shuffled decks.

The questioning is divided into three parts. The first part is operations of arrangements: We put the three parts of either deck on the table and ask the child to construct as many different numbers as he can with two digits (or as many different pairs of railroad cars) with the cards given.

In the second part we shuffle the cards in the sight of the child, and then ask him to draw two of them and to predict which ones he will get. In the course of the drawings, the child makes a list of the elements drawn (railroad cars or numbers).

In the third part of the experiment we analyze with the child the list which he drew up. After having him note the inequalities of the distribution, we have him imagine that there is a basket full of these same cards and ask him if the inequalities will grow larger or smaller with the increase in number of cards to draw from.

The stages observed are the same that we saw in the preceding chapters. Up to about seven or eight years, there is neither systematic arrangement nor understanding of random mixture. There is, however, illusory prediction of isolated draws, that is, an inability to recognize fortuitous arrangements. From seven or eight to eleven years, we observe a beginning of systematizing of the operations, and the beginning of an understanding of chance, but with no generalization of large numbers. At about eleven or twelve years, the operations are done systematically and the fortuitous arrangements are seen as compensating each other in the case of large numbers.

2. *The first stage: Empirical arrangements and failure to understand random mixture.*

The following reactions, when compared with those we know already, present one new aspect: The mixture results from the superposition of the cards and the operations at play are related both to combinations and permutations. It is, therefore, useful to analyze a few examples:

MAST (5; 5) finds by means of locomotives (L), passenger cars (P), and freight cars (F), the pairs PF, PP, LL, and PP. "Are there two the same?" (He takes away PP.) "Any more?" *"PF."* "Can you make any more?" *"PF."* "Haven't you already made that twice?" *"Oh, yes* (he removes both)." "Can you make another?" (He again makes PF.) "Couldn't you make it in the other way?" *"Yes* (FP)." "Any more?" (He remakes PF.)

Drawing lots: He draws FP. "Is there any way of knowing what we will draw next, or not?" *"Yes, LP."* (He draws LL.) "And now, will it be the same or something different?" *"The same thing."* (He draws LL.) "And now?" *"It will be FF."* (He draws PP.) "And now?" *"Again all the passenger cars."*

BER (6; 10) also constructs arrangements in a purely empirical manner. Step by step he gets to eight. "Is that all?" *"Yes."* "Could you find a way to know if you have them all?" *"I don't know."* "Look at this side (first column). How many P's are there?" *"Three."* "And how many freight (F)?" *"Three."* "And locomotives (L)?" *"Two."* "Isn't it strange that there are only two?" *"Don't know."* "And on the other side (third column), how many P, F, and L?" *"Three, three and two."* "Then what could we do?" *"Add another"* (he finds LF). "Do we have them all now?" *"I don't know."*

Random mixture: "Before drawing, do you know what you'll get?" *"Oh, no, not me."* "Do I?" *"Yes."* "Why?" *"Because you have seen it."* (He has just observed the shuffling of the cards.) "But, no. I don't know any more than you do. (We draw a few.) And now, do you think we'll get some we've already drawn, or some others?" *"It will be some of those* (already drawn) *because there are already some there."* (We continue: He gets three PL.) "Does it happen often that we get the same ones all the time?" *"Don't know."* "Will it again be the same ones?" *"I hope not."* (He draws the same ones) "Now a different one?" *"Yes."* "Why?" *"Because the others have some already, and so it wouldn't be right if it were always the same ones."* "Try it." (He draws LF.) *"That's because there was only one of them here."* "And what if I have a lot of cards, a pile like this (gesture), what will it be?" *"If it works as it should, then there will be one of them all over, then two, then three all over."* "And what if it doesn't work as it should?" *"Maybe two there and eight there."*

PIL (6; 1) gets four pairs by gropings. "Is that all?" *"Yes, there are no other ways."* "What could the locomotive also push?" (He finds LP, already done, then LF, and so forth, and gets nine.) "Is that all?" *"Don't know."*

Drawing: same reactions as Mast at first. "And the next one, will it be one of those we already have, or a different one?" *"We'll get one we've already gotten."* "Why?" *"They come out easier. . . . We are going to get LP because we get that easily, we've already gotten it a few times."* "You think there are some that come out more easily?" *"Yes, these three* (LL, PF, LP) *come out more easily than the others."* The next few come out with no regularity, and then, *"We are going to get LP because it comes out more easily."* But after several new drawings, he concedes, *"The last two will be FP, because they come out less easily; we've never had them yet."* "You are sure or not?" *"Yes, I'm sure of still getting FP."* "Why?" *"It's the lid* (of the dealing machine) *that makes them come out."*

MON (6; 6), after making empirical arrangements in the same way as the others, predicts the drawings according to the first observed frequencies. *"Again those two* (LP) *because they are easiest."* "Why is it more difficult to get PL?" *"It's a little easier to get LP than PL because it's reversed."*

JAC (6; 6) counts up to fifty. Using pairs of numbers, he finds the arrangements 12, 23, 33. "Are there any more?" *"I think that's all."* "When you have 23, can't you arrange the digits in another way?" (He inverts it to 32.) "What about with 12?" (He inverts it to 21.) "More?" *"I think I can't do any more."* (We show him 22 and 31.) "More now?" *"I think that we've done them all."*

For the random mixture, he first predicts arbitrarily, then by taking into account the observed frequencies, and finally by compensations. "What does it depend on?" *"I don't know."* "What if I mixed up the 1's, the 2's, and the 3's all together?" *"Then it would come out as 11, 22, and 33."* "And if I mix them only a little?" *"I don't know how that would come out."* "But are there some that would come out more than others?" *"Three's because when we turn 123 around, it is the 3 that is first* (321)."

LUC (7; 3) can count up to one hundred. First the only arrangements that he finds are 13, 33, and 32. "How many two-digit numbers can you make with the three digits 1, 2, and 3?" *"About fifty."* Little by little he discovers six, then eight, and finally nine. "Can we find a way to be sure that we made them all?" *"Yes, make a lot of them and then take out the ones that are the same."*

Random mixture: same reactions. He first thinks that the same pairs turn up by a kind of favor of chance, and then he notes an equalization of frequencies. "If I draw many, many pairs of cards, say one hundred?" *"There will be more differences."*

In examining these operations of arrangement, we see that the subjects of this first stage proceed by gropings, without suspecting a system: The child takes any card, puts it with any other one, and simply looks around next to see if that pair is already on the board. At the beginning of the stage, the permutations present some difficulties. As we have already seen, it is easier to combine new elements than to shift their order, and especially easier than taking into account at the same time combinations and possible permutations.

As for the mixture, the absence of a system in the arrangements is shown, once more, by the failure to understand the effects of mixture and by the belief in a hidden order in the shuffled cards. This hidden order which the drawings reveal is sometimes based on the observed repetitions and sometimes on compensations which are of a moral type (as Ber says, *"It wouldn't be right"*). As for predictions founded on the given frequencies, it may be a question of a feeling of empirical repetition (Mast) or because of the idea that the cards already frequently drawn *"come out more easily* (Mon and Pil)." With the numbers the child sometimes allows himself curious predictions which give eloquent witness of his inability to understand random mixture. For instance, Jac, thinks that the digits 1, 2, and 3 will always produce 11, 22, and 33. And Jac also thinks an insufficient mixture will give unpredictable results, except that the 3 will show more frequently because *"when we turn the number around, the three is first."* Therefore, Jac thinks an incomplete mixture is the equivalent of a permutation.

Thus we understand the reason why, when the subject lacks a combinatoric system, which is similar to an inadequate idea of random mixture, he constructs his own arrangements haphazardly, while denying the role of chance in the arrangements which are, in fact, fortuitous. In both cases the child lacks the understanding of the large number of possibilities, and this is why, in the first of the cases, he makes only a few trials; and in the second he believes that there is an order hidden in the observed frequencies:

3. *The second stage: Search for a system and the beginning of the idea of chance.*

As was the case with combinations and permutations, during this second stage we see a progressive series of tries at a systematizing of arrangements, and at the same time, a more and more correct interpretation of the fortuitous arrangements constituted by the mixture. Here are several examples, beginning with a case intermediary between stages I and II:

BON (6; 10), one after the other, makes 13, 12, 11, 32, 23, 31,

22, and 33. "Can you make any more?" "*Yes* (and he makes 21, and then 31 which he removes because it was already there)." "More? . . . What do you have to do to know that you have not forgotten any?" "*I don't know.*" "How many do you have in all?" "*Nine.*" "Can we make any more?" "*No.*" "Could you arrange them better?" (After a hesitation, he puts them in an order: 12, 13, 11, 32, 31, 33, 23, 22, 21.) "Can you do it even better than that?" "*Yes, like this* (11, 12, 13, 31, 32, 33, 21, 22, 23)."

Mixture: After several tries he predicts that the most likely pairs to be drawn will be those that have not yet shown "*because the same ones don't always come out* (since there is a mixture)." Then, after noting some repetitions, he continues to predict that the pairs not yet seen have less chance of being drawn: "What is most likely to be gotten this time?" "*What we've already had because there are more of them than of those we've not gotten.*" After this he refuses to make any more predictions, even concerning the final distribution of twenty pairs and especially for one hundred pairs.

DES (7; 10) immediately constructs 11, 21, 23, 31, 32, and then hesitates. "Do you know how many there will be in all?" "*There will be maybe* . . . (he adds 22, 3 . . .) eight like that. *Yes, I see I can make 33 too. I counted those I made. Oh, yes, we can still put 12.*" "What can we do to make sure?" "*I look at the numbers.*" "Are you sure just by looking? . . . Do them over again putting them in order." "*There, 11, 12, 13. Oh, and I forgot 21, 23. And 22 is missing. And 31, 32, 33. That makes nine.*" "Can you do more?" "*No, we can't do any more.*" "What if I let you use the digit 4?" "*Then I could make 41, 42, 43, 44.*" "Those are the only ones?" "*We could put the 4 in different positions: 14, 24, 34.*" "That would make how many numbers?" Des does not yet know the multiplication table, but by successive additions he finds sixteen arrangements. With a similar procedure, he is even able to find twenty-five two-digit arrangements using five different digits.

MET (8; 10): "With these three digits, how many different numbers can you make of two digits each?" "*Three, no, more than that. Six, I believe.*" (He makes one after the other 12, 21, 33, 11, 22, 23, 32, 13 and 31, and keeps looking around to see if he has made them all.) "Are there any more?" *"Yes, a few still."* "If you think about it, can you know if you made all there are?" *"No, oh, yes, I can look to see if there are 1, 2, and 3* (in equal number), *and if we made all the pairs."* He counts then all the 1's, 2's, and 3's. *"Yes, that's all."*

Random mixture: "Can we make a guess?" *"No."* "Do you think that 12 will come out first?" *"No, since they come out by chance."* (He gets 23.) "Can you tell me why 23 came out? Is there a reason?" *"There is no reason for getting 23 because you shuffled them."* Next he believes that the ones that will most likely come out are *"those which haven't yet come out."* In the same way, he says about the pairs that have already been drawn, *"I almost think that they won't come out any more because they've come out so often already."* Since the final result does not confirm this idea of compensation, that is, is not regular, Met concludes that if we do it over again *"it will be quite different."* But we cannot predict with certainty *"because we shuffled all the cards. Not even you can know."* There is no intuition of the law of large numbers.

Web (9; 6) makes 21, 31, 22, 33, 12, and then makes 13 by reversing 31. *"I can't make any more. Oh, I see one that is to be made—11* (looking at 22 and 33 which he made)." "How do you know that you have them all?" *"Since we have three digits, each digit occurs three times."* "That makes how many?" *"Nine, but I only have seven."* "How can you find a way to be sure?" *"By looking."* "How many 1's are there in the numbers you have found?" *"Six* (without counting)." "And 2's?" *"Three."* "And 3's?" *"Three."* "Look." *"Oh, six times for each one* (he adds 32 and 23). *Oh, I see. If there weren't the same number for each, it wouldn't be right."*

Random mixture: *"It is less likely that we get two digits together* (e.g., 22) *because they are too well mixed."* "What if we had mixed them only a little?" *"If it's only a little, then it's likely that they won't be together."* Toward the end of the experiment, 13 has not yet come out and he predicts it. "Are you certain of getting 13?" *"It depends on how it is mixed. If there were more cards, it would be more unusual not to get 13."* But he does not yet generalize to large numbers.

DUM (10; 8): "How many pairs of numbers do you think you can make with the digits 1, 2, and 3?" *"Eighty."* "O.K., take a try." He gets eight. "What can you do to be sure you have them all?" *"Taking 3, 2, 1 and trying to make pairs, and then looking to see if you already have them* (he puts his eight pairs into order)." "How many do you have starting with 1?" *"Three."* "And with 2?" *"Three."* "And with 3?" *"Oh, yes. I've got to make another one with 3."* (He finds it.) "Is there a better order to put them in?" He puts 11, 12, and 13, one below the other, and then 21, 22, 23, and 31, 32, and 33.

Mixture: *"It's mixed up, and so we can't know what will come out."* Toward the end he predicts the digits that haven't been drawn *"because we should have about the same number of all."* It could happen, however, that some numbers don't come out; *"that is not so certain* (as is the regularity)." Thus there is an understanding of the law of large numbers, but for small collections and without any generalization for true large numbers. "If we did it with several piles like that one, would it be more or less regular than just using this one?" *"It would be more regular. Oh, no, it would be more different if there were many cards."*

MOR (10; 8): Same reactions at first. As for large numbers, the differences would stay the same: *"It will be the same thing because we play it often."*

In these operations of arrangements, we find in the subjects the same progressive sense of regularity and the same gradual systematization that we have already observed in their experiments with combinations and permutations. Web and Dun, for example, are immediately certain that if we find *n* arrangements beginning with 1 and with 2, then it will also be necessary for there to be *n* arrangements beginning with 3. In the same way, when the child has found 11 and 22, he generalizes immediately to 33. As for the method of discovery of the different possible arrangements, it always remains empirical at the beginning, but the subjects at this stage quickly understand that the combinations found can be ordered according to the first numbers (1, 2, or 3) in each arrangement. It is only after doing this that they have the idea of putting the second number in the same order (11, 12, 13). Certain subjects even succeed empirically in understanding the law according to which the number of arrangements in pairs for three elements is 3^2. But this is only an empirical discovery (in particular, for Des who does not know how to multiply, but in spite of that comes to a generalization for four and five digits), and the discovery is facilitated by the fact that the elements to be arranged are numbers, which by their nature are able to be ordered.

As for random mixture and the idea of chance, we find in correlation with the gradual systematization of the combinatoric operations an understanding of the fortuitous character of the arrangements created from the mixture. This is seen when Met declares immediately that if 23 is drawn rather than another number *"there is no reason for this since the cards were shuffled."* But if the isolated cases are from then on understood as unpredictable, the child recognizes as well that the distribution of the whole tends to be regular: *"We will have about the same number* (frequency) *everywhere,"* Dum says. However, this is a case of understanding which is limited only to what we have earlier called small large numbers, that is, with sets of objects which can be seen and handled. As soon as we try to have them generalize the law to include true large numbers, the child resists: *"It would be more different if there were many cards,"* says Dum, for example (and he is already ten years and eight months). But he lacks the ability of formal deduction and the feeling for proportions. Mor, at the edge of stage III, says even that *"it would be the same thing,"* but without admitting, however, that the compensations continue to increase with the number of drawings.

4. *The third stage: Understanding of the system of arrangements and of the laws of random mixture tending toward large numbers.*

It is interesting to note that, again in this experiment, the age of eleven to twelve marks a turning point in the evolution of the ideas of chance as well as in the understanding of combinatoric operations. But for arrangements, as for permutations (which they imply), it is correct to distinguish two successive levels during stage III: a substage III A in the course of which the child succeeds in discovering, for a given number of elements (*n* equals three or four), the law of the square (arrangements in pairs being thus n^2, if we include in it the repetition of a number with itself such as 11). But he does not understand the reason for this, and he does not, therefore, succeed in making a constructive generalization from one number of elements to another, but only an empirical generalization, and then only in certain cases. At level III B, on the other hand, the reason is understood and inspires a generalization.

GER (10; 8): We begin with only two elements, 1 and 2. Ger finds immediately 12, 21, 11, and 22: "How many?" *"Four."* "How do you know that's all?" *"There are two different digits which can be reversed* (12 and 21)." "And can you say what we will get with three elements before we try?" *"Six."* "Why?" *"There are three different digits and we can invert them two times."* "Try it." *"There.* (He makes 13, 31, 12, 21, 32, 23, 11, 22, and 33.) *That makes nine."* "Can we put them in a different arrangement?" *"Yes* (he orders them differently based on series of inversions)." "Can you say ahead of time what we will get with four digits?" *"Yes, sixteen."* "How did you figure it?" *"By multiplying four times four. We have four digits and we can arrange them in four different ways."* "How did you get that idea (in reality it is by analogy with three times three equals nine)?" (He calculates.) *"It will give twelve numbers* (12, 21, 13, 31, 14, 41, 23, 32, 24, 42, 11, 22, 33, 44)." "Is that all?" *"Oh, no; 43 and 34. That makes twelve plus the four* (11, 22, 33, 44), *and that makes sixteen."* "What if we have five digits?" *"It will be thirty, because there are five numbers, five possible inversions, then five more."* "Try it." He finds twenty-five pairs, but does not understand why.

For the random mixture he refuses to predict individual cases because *"the cards are shuffled."* As for a particular number *"it can happen that it will not appear,"* but when asked, "what will we have when we have drawn them all?" *"There will be fewer differences"* between the frequencies of the drawing for each number: *"It is not certain, but we will have just about the same number of times for all of them."* "What if we had three times as many cards, what differences would we have?" *"There would be fewer. If there are very many, it will become equal."*

ALA (12; 7) shows from the beginning the four possible pairs of two digits. "And what will you get with 1, 2, and 3?" He makes 11, 22, 33, 12, 13, 21, 23 . . . "How many will it be in all?" *"Twelve."* "Why?" *"A series like that with 1 and with 2 etc., and three times four is twelve. Oh, no, eighteen, three piles of six, then that makes nine numbers."* "Now try it with four digits." He makes 11, 12, 13, 14, 22, 23, 21, 24, 33, 31, 32, 34, 44, 43, 42, 41. "Is that all?" *"Yes, I started with all the numbers beginning with 1, then beginning with 2, etc."* "And with five numbers, what will it be?" *"Twenty."* "Why?" *"Five numbers in each column, and four columns. Oh, no, twenty-five because of the fifth column."* "And with six piles, what?" *"Thirty-six."* "Why?" *"It will again be a column and two 6's in each column* (he understands by that 15, 16, 25, 26, 35, 36, etc.) *That makes six times six equals thirty-six."* "Good, and if you have seven piles?" *"Forty-nine numbers."* "Eight digits to work with?" *"Sixty-four. Now I get it. It is because we multiply the number by itself."* "Yes, but why?" *"With each one we put one of each. With six we can put six each time; and with seven, we can put seven each time."*

Reactions to random drawings: same as the preceding subjects.

We see thus that, apart from the final discovery of Ala which leads us to the threshold of level III B, the subjects certainly do discover the law of the square, but without finding the reason for it. It is thus a question of a progressive discovery, but becoming more and more complete thanks to a system used to make successive arrangements. It is the complete system that implies the operations of a second power which characterizes formal thought. But, simply, the ability to use it does not, as we have noted, lead to an understanding of the operative reasons which are at play. Thus Ger still separates the inversions (12 and 21) from the doublets (11 or 22) without understanding that it is a question of a single principle of construction.

The subjects at level III B quickly derive the generalizing principle from their first trials:

MAT (12; 11): With two digits: *"Can I take two counters from the same pile?"* "Yes." *"O.K., that makes 11, 12, 21, and 22."* "What will it make with three digits?" *"It will give six numbers."* "Why six?" *"First we can make the numbers 11, 22, and 33, then*

12, and 13, 21, 23, 31 and 32. That makes nine possibilities." "Are there any more?" "*No, we can't arrange them in any other ways.*" "With four numbers?" "*There will be twelve.*" "Why?" "*Because four times four makes twelve. Oh, no, sixteen.*" "Why four times four?" "*With three counters, there are nine possibilities; thus with four, there are sixteen.*" "But why?" "*Each digit can go with each of the four digits. The 1 with the 1, with the 2, with the 3, and with the 4, etc.*" "Good, and with six numbers?" "*It will be thirty-six, and with seven, it will be forty-nine.*"

GRA (13; 3) finds empirically the nine arrangements of three digits in pairs. "What if I give you four digits?" *I would have sixteen.*" "Why?" (He laughs.) "*Since with one digit and these three others we can make four combinations and then the same thing with the second digit, etc., that will make sixteen in all.*" "And with six digits?" "*There will be thirty-six because with the one pile and the five others, we have six combinations and there will be six piles.*" "And with seven digits?" "*Forty-nine.*"

At this last level, the discovery of the law is accompanied almost immediately by an understanding of the reason for the law. It is clear that this understanding itself is only a conscious grasping of the relationships inherent in the system adopted by the subjects to exhaust all the possible arrangements. In other words, this system, while reaching a completeness at level III A, does not yet bring about a scheme which is, at the same time, reflective and anticipatory; the progress of level III B consists in drawing from this formal system its actual formulation in the manner of propositional statements which describe the constructive principle and which determine the possible generalizations.

5. *The quantification of probabilities based on arrangements.*[1]

We take a sack containing a mixture of twenty red marbles and twenty blue ones, and from it successively extract pairs. Knowing the possible arrangements (RR, RB, BR, and BB), will the subject

[1] Done in collaboration with Gaby Ascoli.

understand that the most probable distribution of the pairs he gets will be five pairs of RR, five pairs of BB, and ten mixed pairs of RB and BR? In other words, will he have the intuition of a simple probability corresponding to the one which the famous law of Mendel, for example, expresses? It seems of interest to us to ask this question to about fifty subjects who are questioned first on the operations of arrangements in such a way as to allow us to study an example of quantitative probability based on these operations and not simply the qualitative probabilities which have been treated in the preceding sections dealing with drawing cards.

First we have the child note the equal number of the two sets of blue marbles and red marbles which we then put in a sack and shake up. Three empty boxes are made available to receive the pairs of red marbles, the pairs of blue marbles and the pairs of mixed ones (which we will note as M). The question is then to know if, after having drawn all the marbles two at a time, there will be at the end more pairs of reds, of blues, or the mixed pairs. We can ask questions either at the beginning about the final results, or we can have the subjects predict ahead of time what each drawing will give. We are looking especially for an explanation of why there will be the particular distribution.

The following reactions in this case correspond to the three stages usually found. In the course of the first stage the child shows a clear tendency to believe that the pairs RR or BB will come out more frequently than the mixed pairs. In the course of stage II, this belief is quickly reversed, but without numerical quantification of the largest frequency of M. Finally, in stage III, the law is found in its quantitative form. Here are some examples of the first stage:

LAN (4; 10) constantly predicts that he will get RR and BB, and not mixed pairs M, even though it is these mixed pairs which we obtained by chance for the first five trials. Sixth drawing: "What now?" *"It will be red and blue."* "Why? . . . Take a try." (He gets BB.) "Now what?" *"Two red ones."* (He continues to predict red pairs or blue pairs to the very end.)

VAT (5; 6): "What will they be?" *"Two blues or two reds."* "Could you get anything else?" *"No."* "We'll see. (RR) What now?" *"Two blues because we already have two red ones."*

"Draw." (He gets a mixed pair and is astonished.) "You didn't think about that one?" *"No."* "And this time?" *"Don't know."*

UL (5; 9) immediately predicts RR and BB in the same way without thinking about the mixed pair M, and then is quickly proved wrong by the experiment, but concludes nothing from this. "If we had emptied the whole sack, what would we have gotten most often—reds, blues, or mixed pairs?" *"I have no idea."* "Guess." *"Reds or blues."*

These reactions which conform to all the traits which we have already learned reveal an interesting attitude of the little children regarding random mixture: The unmixed pairs RR and BB seem to have for them a higher probability than the mixed pairs M. More precisely, the idea that there could be a mixed pair does not even seem to strike them, in spite of the fact that we did the mixing right in front of them. It is as if the spontaneous arrangements produced in the sack were supposed to lead to a union of the similar elements rather than to a mixture.

But in the second stage the possibility of the mixed pairs is seen from the beginning, and its probability quickly judged as higher than the probability of the unmixed pairs (RR or BB).

ARL (8; 10): "You can take two of them. What color will they be?" *"Reds or blues, or a red one and a blue one."* "Which kind will be more likely?" *"Maybe the red ones because maybe they don't get as mixed up as the others."* We shake the bag again and Arl gets two red marbles. "What now?" *"It could be something else since it is mixed up: two blue ones or a red and a blue."* After a few more drawings: "What this time?" *"The reds and blues mixed come out more quickly."* "Why?" *"Because the marbles are better mixed."* "If we do it again several times, will it be the same thing?" *"No, it's chance."* "With one hundred marbles?" *"The mixed pairs will be easier to get."* "With forty marbles, how many mixed pairs will we get?" *"Don't know, but more than we do of the others."*

JAR (9; 4). *"Red ones or blue ones or mixed red and blue."* "Which is more likely?" *"We can't know."* After several drawings: *"We get the mixed ones more often."* "Why?" *"Becaues it is more unusual to get two reds or two blues together."*

ROL (9; 4) predicts two blues on the first drawing (which is confirmed) then a red and a blue (which is also confirmed). "And what are we most likely to get next?" *"Reds and blues, the mixed are easier because it is mixed by chance."* "Why? *"Somtimes the mixed ones come out, and sometimes not, but it is rarer to get two red ones or two blue ones."* The experiment confirms this little by little. *"Yes, maybe the box of mixed ones will always be fuller because all of them are mixed in the sack so that we more often get a mixture."*

DEL (9; 10): First drawing: "What will we get?" *"Maybe two of the same color, maybe one of each color."* "What will we get most often?" *"One of each color because we mixed them, and they aren't all in the same place."* "With forty marbles, how many pairs will be mixed?" *"I don't know, but more of them."*

Thus we see that the possibility of mixed pairs is seen from the beginning, and often understood from the first trial as having a greater frequency. From the first drawings this probability is affirmed or increased in the eyes of the child, but without a numerical quantification.

The law is formulated in the course of the third stage:

GRIV (12; 5) predicts before any drawing that he will get: *"Two red ones, two blue ones, or a red one and a blue one."* "Which will be most likely?" *"It's the same. We don't see what we are getting. Yes, the mixed ones because there are half blues and half red ones. We can get the mixed ones twenty times, but blue pairs only ten times."* After several drawings: *"There will always be more mixed ones."* "Why?" *"Because there are half of each, and we have a better chance of getting red and blue together."*

MAR (12; 6) before drawing any: *"We will get two blues, or two reds, or a red one and a blue one."* "Which most often?" *"Blue and red, I believe because it is mixed."* "Why?" *"Because it is chance that mixes them. We will have mostly mixed ones."* "With twenty pairs?" *"Maybe ten mixed ones."* "And the others?" *"Five for each one."* "Are you sure?" *"We could get something slightly different, but less often."*

GAN (12; 6): *"There will always be more mixed ones."* "How

many?" *"A half maybe; ten pairs and then five pairs of red ones and five pairs of blue."*

KONJ (13; 3): *"More likely the mixed ones."* "Why?" *"Because we put forty marbles in, and thus we have a better chance of getting the mixed ones, one-half the chances."* "Could we get always mixed ones?" *"It would be rather strange."* "What if we play a large number of times?" *"Ten mixed pairs, five reds and five blues."*

We see how the question of probability is rapidly quantified according to the proportions five RR plus five BB plus ten M (RB or BR) in the course of this third stage in spite of the absence of any explicit allusion to arrangements. This is all the more striking because, without having distinguished the relationships of order BR or RB the subjects anticipate exactly the numerical relationships at play, instead of attributing a one-third probability to each of the categories (RR, BB, and M). This is a new example of the structuring role that combinatoric operations play concerning the notions of chance and probability.

X. Conclusion: Chance, Probability, and Operations

Logical and arithmetical operations constitute systems of actions interrelated in a rigorous way and always reversible, this reversible aspect rendering deduction possible. Transformations occurring by chance, on the contrary, are not interrelated in a rigorous way and the most probable systems that they form are essentially irreversible. While logical and arithmetical operations lead to and end in the construction of groupings or groups fortuitous transformations remain nonreducible to this type of structure. Finally, induction lies between operative deduction and nondeducible fortuitous transformations: induction comprising those steps by which one sifts through experimentation separating what is fortuitous from what is deducible, while at the same time preparing for deduction itself.

To interpret the different results obtained in this study and come to a conclusion, it is proper to discuss these results in relation to the whole of the results furnished by our studies of the operative development of intelligence. As we have seen constantly by contrast with operations, chance is gradually discovered, and it is by constant reference to the structures of operations that chance is finally understood and yields a system of probabilities.

1. *The three stages of the development of the idea of chance.*

A first fact is immediately striking when considering the reactions of the children as a function of their age: This is the direct correlation between the formation of the notions of chance and probability and the formation of the different operations whose psychological genesis we are studying.

From the point of view of the genesis of the elementary logical-arithmetical operations (groupings of classes and relationships and sets of whole and fractional numbers), we can, in fact, distinguish three major stages of development. The first is before seven or eight years of age and is characterized by the absence of what are properly called operations, that is, a type of reversible composition. The reasoning used at this stage remains prelogical and is ruled only by systems of intuitive regulations, without hierarchic nestings, without conservation of totalities, without rigor in possible inferences. But there is a progressive articulation of intuitive relationships leading little by little to an operative level. From seven or eight to eleven or twelve years of age, a second period is characterized by the construction of operative groupings of a logical order, and by numerical sets, but on an essentially concrete plane, that is, related to objects which are able to be seen or handled in their actual relationships. Finally, a third period begins at eleven or twelve years of age characterized by formal thought, that is, by the possibility of tying together one or several systems of concrete operations at the same time, and in translating them into terms of their hypothetical-deductive implications, that is, into terms of the logic of their propositions.

It is in this way that we explain the evolution of the combinatoric operations for which Chapters VII to IX of this work have given us the necessary indications. During the first period (before seven to eight years) the child does not even suspect the possibility of a system which would permit him to find all the combinations of pairs, without missing a single one, and all the permutations or all the arrangements in pairs which could be made by means of the several elements. There is nothing surprising in this failure, since what is needed are special multiplicative operations, and the subjects are not yet even in possession of additive or simple multiplicative operations. In the course of the second period (seven to eleven years), the child understands the possibility of such systems but discovers them only in an empirical and incomplete fashion, since these systems suppose formal thought by their conjunction of several concrete operations at the same time. Finally, at the level of the third period (after eleven or twelve years), the beginnings of formal thought allow the discovery of some complete combinatoric systems for a small number of elements.

This double operative development converges in a rather precise manner with what we have seen in the study of the notions of chance and probability. Whatever, in fact, may be the experimental apparatus adopted to shed light on the reactions of the child concerning chance, we have constantly found the existence of three successive periods corresponding closely with those of the evolution of logical-arithmetical operations and combinatoric operations.

During the first of these three periods (before seven or eight years), the child does not distinguish the possible from the necessary and moves, therefore, in a sphere of action as distant from chance as from operation itself. His thought oscillates between the predictable and the unforeseen, but nothing for him is either predictable for certain (by which we mean deriving from necessity), or absolutely unforeseen (that is, fortuitous). Thus we could not consider his anticipations as judgments of probability, deriving from a greater or lesser degree of subjective certitude, because this certitude is only the product of a failure to differentiate between practical notions of intuitive predictability and caprice. At this age then there is neither chance nor probability, because the system of reference based on deductive operations is lacking.

At about seven or eight years of age logical-arithmetical operations appear and this begins a second period which marks the first development of the idea of chance. On the one hand, in fact, the discovery of deductive or operative necessity allows the subject, by antithesis, to conceive of the nondeducible character of isolated and fortuitous transformations,and to differentiate in this way between the necessary and simply the possible. On the other hand, the operative union of the complementary parts into a whole (A and A′ into B, for example) leads to concrete disjunction: [1] If x is an element in B, it can be in A or in A′; and by this very fact, this concrete disjunction entails the notion of double or multiple possibilities which are implied in all judgments of probability. In brief, the discovery of the existence of chance, in fact, results from contrast with the construction of operative necessities, and the beginning of the structuring of the relationships of probability is due to the possibility of conceiving of the sets of mixed elements as totalities formed from quantifiable parts. Moreover, mixture as such, that is, the simplest physical expression of chance, begins to be conceived of as a real mixture and no longer as only an apparent disorder veiling an underlying order. The reason for this is that the child succeeds in seeing through to the possibility of fortuitous combinations, permutations, or arrangements (which, in fact, are conditions of all mixtures), while at the first stage the sight of the actual mixture did not shake the convictions of the subjects in an imaginary order excluding chance, and this was true because of their failure to understand combinations or possible permutations.

But it is only during the third stage (after eleven or twelve years) that the judgment of probability becomes organized in its general aspects by a kind of ricochet of operations on chance. From this point of view, we can characterize this third stage in the following manner by comparing it with the first two stages: During the first period there is no differentiation between what is deducible and what is not because the intuitive anticipation remains halfway between operation and chance itself. During the second period there is differentiation and hence an antithesis between chance and operations, these latter determining the domain of the deducible while chance defines, therefore, the domain of the incomposable and the irreversible, in a word, the unpredictable (cf. in this regard the phase of doubt which often characterizes the beginnings of stage II: see for example Chapter III). In the course of stage III, on the contrary, there is a synthesis between chance

[1] That is, based on the existence of separate or disjointed classes A and A′ and not yet on the logic of the propositions.

and operations, the latter allowing the field of fortuitous distributions to be structured in a system of probabilities by a sort of analogous assimilation of the fortuitous with the operative. Two correlative processes come together to achieve this result. On the one hand, the construction of combinatoric systems (marked by the discovery of a method permitting one to construct all the possible operations using a small number of elements) leads the subject to conceive of a random mixture as the result of transformations, but executed without order and realizing only a part of the total possibilities. Formal thought, on the other hand, which allows the construction of such combinatoric systems leads also to the discovery of proportions. The law of large numbers, which applies the relationships of proportionality to these same combinatoric operations, leads the subject to conceive of the legitimacy of a probabilistic composition of the fortuitous modifications in the sense of a distribution which becomes proportionally more regular and, consequently, understandable in its totality, if not in the details, as a rational prediction.

Thus, individual evolution of the idea of chance is achieved and probabilities based on large numbers indicate a kind of synthesis between operations and the fortuitous. This synthesis appears after the antithesis, which was radical at the beginning of the second stage, as well as marked by a failure to differentiate, which was characteristic of the first stage.

2. *The first period: Failure to differentiate between the possible and the necessary.*

The most general characteristic of the first of these three periods whose principal traits we have just sketched is, no doubt, the one which distinguishes the modality of judgments proper to this mental level. Observing things superficially, we could have the impression that the young child, and even the baby, dissociate the possible from the necessary. When the nursing child hears a noise behind the door and expects, without being sure of it, to see his mother appear, we could say, in fact, that he considers this appearance as possible and not as certain, or even as probable to some degree. Such an interpretation would lead us to conceive of the judgment of probability as very primitive and even anterior to precise ideas about chance, or as related to intuitive notions of the uncertain or the unexpected.

Here, as everywhere else, we must know how to avoid the illusion of the implicit which consists in attributing the most complex notions to elementary levels as if they were already contained in the initial intuitions, and the illusion of the discontinuous which may lead one to admitting creations *ex nihilo* when they are developing. Functional continuity and structural transformations are, in this case as in all others, the two basic conditions of genesis. It is thus that all animals, considered biologically, feed and present, consequently, an invariant function of nutrition, but this function is exercised by organs whose structures differ and it would be absurd to try to find in the protozoa the analogy of a stomach or the small intestine. Psychologically, in the same way, the sensorimotor adaptations of the first year present the same functions of coordination and generalization as does logical thought; and one can thus compare functionally the sensorimotor schemes to some sorts of practical concepts and their reciprocal assimilations with some sorts of judgments or even reasoning. But it would be idle to draw from these functional analogies a structural identity and to attribute to the nursing infant operative structures, whether deductive or probabilistic.

At the sensorimotor levels of intelligence, which are at this time preconceptual and preliminary to true thought, the subject is certainly capable of intuitions based on practical assimilations or diverse representations, and he knows quite well that his anticipations are at times incorrect because reality is not always in conformity with what we expect of it. For the empiricism of sensorimotor intelligence and of preoperative intuition, the real remains, therefore, midway between the regularities (due to assimilation) and the unexpected which constrains the anticipatory schemes to various accommodations. But in the same way that regularities intuitively perceived do not yet allow a grasp of the necessary logical operations, so also the unexpected (which the child notes simply as a contrast with the expected regularities) is not to be confused with the unpredictable, that is, with the modifications conceived of as not reducible to possible deduction. The unexpected is initially, in fact, conceived of as only the expression of fantasy or of the arbitrary and is, therefore, interpreted as a psychological or intentional mode of motivation, and not yet as a logical or physical mode of the noncomposable and irreversible.

Concerning the modality of judgment, the subject could, consequently, not know the distinction between the possible and the necessary (and the same would be true during the second period) precisely because he lacks the notion of necessity inherent in operative deduction. From the functional point of view, there is certainly at this time a notion which performs the function of the possible, and this is precisely the idea that the near future is made up of events which one is not certain that he can anticipate. But from the point of view of structure, this uncertainty does not have the same structural value as a logical unpredictability, that is, as a notion well differentiated from necessity. As long as the subject is not certain that A will equal $B - A'$, and A' will equal $B - A$ if we have $A \cup A' = B$ (and this certainty is not acquired until about seven years) [1] we cannot call possibility the quality deriving from his uncertainty, when working with the mixture of two elements X and Y, he wants to know which of the two he will draw. In fact, if there is not a feeling of necessity in the first case, the lack of certainty in the second one remains undifferentiated, which means that in both cases there is still the failure to differentiate between necessity and possibility.

In short, we can characterize the first stage in a general way by the intuitions of the actual which present a whole spectrum of variations lying between motivation and nonmotivation. The first ones thus announce the deductive operations and the second ones chance, but they remain dissociated from each other viewed in their structures because of the lack of any operative construction of the logical unions. We understand then the reasons for the diverse characteristics we have been led to attribute to this first stage in the course of the preceding chapters.

In the first place, the child does not suspect the true nature of

[1] The same is true of certainties of the type $A = C$ if $A = B$ and $B = C$, etc.

random mixture and tries constantly to find within the disorder, which he considers as only apparent, a hidden order of some kind based on the common properties of the elements, on their arrangements before the mixture takes place, and so on. This systematic reaction which we noted principally in Chapters I and V is really only strange and incomprehensible for those minds already in possession of deductive order, because then the mixture appears to be precisely the contrary of such an order. But for the subjects still incapable of what is properly called deduction, it is natural for disorder to be able to mask a hidden semiorder since deductive order itself is a result, as they see it, of such analogous intuitions. Instead of composing operations when they are trying to operate deductively, they, in fact, know only how to guess at a hidden order, which is consequently only a partial order. In both cases the child of this first stage succeeds only in calculating intuitively an approximate order, where older subjects see how to contrast a deducible order with a disorder that resists any deduction.

In the second place, irreversibility which is characteristic of fortuitous processes, and especially that irreversibility which characterizes random mixtures, is not understood by the child at the first stage, and for the same reason. They have not yet discovered operative reversibility, and lacking such a system of reference, they consequently would not know how to take cognizance of the irreversibility which additive noncomposition in the nondeductability of uncertain and fortuitous processes entails. In fact, they admit with no difficulty the return of the pieces in the mixture to their starting points, that is, an unmixing, and, as we have seen in Chapter I, where it was not a question of true reversibility, but simply the idea of an initial order which has been momentarily thrown into disorder by an accident and which has the tendency to reassert itself again, since in reality it has not ceased to be an agent.

In the third place, when it is a question of predicting isolated fortuitous cases, the reactions of the subjects are not based simply on resemblances, initial order, and so on, but also on a relationship which comes up frequently and which, at first glance, seems to be of a probabilistic nature: compensation. If, for example, we have two eventualities, A and B, and A appears more frequently, the subject will then bet on B because it has been skipped until then. But, as we have seen also, this is not at first a situation of statistical compensation of the type which leads to progressive regularities and to the law of large numbers. It is, rather, an intuitive notion explainable, as were the preceding ones, by the preoperative character of this stage, and which, like all intuitive notions, includes in its undifferentiated state certain subjective or egocentric liaisons joined with certain objective observations. Its meaning is something like *"each one gets his turn,"* as in games, sharing, and social conduct based on justice, and also like physical alternations such as good and bad weather. Therefore, this compensation is the manifestation of an order of things antedating the idea of chance, and not a probabilistic composition of events which, in so far as they are fortuitous, may be as likely to happen in one way as in another.

In the fourth place, and as a proof of what we have been developing, in the case of a distribution centered on a median value such as the bell curve, the child at the first stage does not suspect the necessity of a symmetry between the opposed values (see Chapter II). This symmetry is precisely the expression of a compensation which would be of a statistical order, as contrasted with the compensation expressed by *"each one gets his turn."*

In the fifth place, when the child is not successful in predicting isolated fortuitous cases by basing his answers on compensation, he frequently begins basing his predictions on the largest frequencies observed up to that time. In the case of two eventualities, A and B, the subject will choose A if it has been drawn more frequently than B. If no objective reason allows us to believe that A is, in fact, more likely, it is clear that we have here again the belief in a hidden order which is manifesting itself by the frequency.

In the sixth place, from the structural point of view we could not relate these reasonings based on frequency, made by a subject of the first stage, with inductive reasoning even if the generalization that they show announces the generalization that will come into play with induction. In fact, what is properly called induction supposes at the same time a knowledge of deductive operations and of chance itself, since inductive reasoning consists precisely in sifting what is regular from what is fortuitous, organizing the regularities into a system of classes and relationships able to be treated deductively. Consequently, when we ask the child of this first stage about problems whose solution implies a dissociation between inductive regularities and fortuitous distributions (such as the rotation of a needle with or without magnets affecting its stopping points, as in Chapter III), we note that the children do not look for this dissociation. Because they are convinced that the needle does not stop at fortuitous points and that, therefore, they must try to account for the actions by noting resemblances between colors of the points of stopping, the order and succession of the colors, and the like, they interpret the controlled stopping of the needle in front of the magnetized boxes by the color of the boxes, and so forth, or they do not even try to explain this regularity at all, seeing in it nothing more than what for us would be facts attributable to chance. To explain this, we could say that everything at this stage is still inductive, since there is neither chance nor deduction but simply an intuition of the real or imaginary regularities. It is clear, however, that at this time the term *induction* would lose all meaning and would become confused with simple prelogical or egocentric generalization.

In the seventh place, because the child lacks operative deduction and any notion of chance, he is not at all surprised by miracles such as the one when all of the counters land with the cross showing when there is supposed to be a cross on one side and a circle on the other (Chapter IV). The reason for this is that, since they lack deductive operations and the notion of chance, everything for them is, in differing degrees, miraculous. It is only when we have a rational mind, that the question of miracle is posed because it is contrary both to natural regularities and to fortuitous fluctuations. For the ancients, on the contrary, miracle (in the etymological meaning of "marvel" or "wonder") was a natural thing, and for the primitive, everything is, in fact, a miracle. It is precisely the same reaction which our subjects of the first stage present, and for the same reasons.

In the eighth place, the absence of operative unions is shown in the clearest manner in the question about quantification of chance, thus about probabilities in the strict sense (Chapters V and VI). When given a set B formed of a single element A and two elements A′, it happens rather frequently that the child judges that he will be more certain of drawing an A, if he is drawing a single element, because the drawing is for a single element and A is the single element in the mixture. We have already seen that such behavior is explained only by the failure to make a relationship between the parts and the whole. Because he fails to make the inclusive addition $A \cup A' = B$, the subject does not say to himself that a B can be either A or A′ and that in such a case the eventuality of getting A′ is greater because the part A′ is represented by two elements. On the contrary, he considers the part A as being given independently, and he chooses it without consideration of A′ or of the whole set B. If the absence of operations explains, therefore, the failure to understand chance as a whole group of noncomposable and irreversible events, the absence of operative unions explains in its way that the predictions of the child do not yet present any probabilistic structure which can relate the part (favorable cases) with the whole (possible cases). When quantity becomes a part of the judgments of the child, it is then a question of a particular factor among others, but there is not yet any quantification of the prediction as such, and thus no judgment of mathematical probability.

In the ninth place, and for even stronger reasons, we do not yet find at this stage any reasoning related to the whole field of distributions (large numbers, etc.).

Taken altogether, the principal characteristics of this first period certainly do constitute a homogeneous whole, explainable in all its details by the preoperative level of the child as well as by the undifferentiated nature of his intuitions based here, as in other domains, on his proper and immediate activity and not yet in a coordination of the actions in reversible operations.

3. *The second period: Discovery of chance as a noncomposable reality as an antithesis of operations.*

Between six and eight years, and at a median of seven years, the spatial-temporal and logical-arithmetical intuitions, after showing progressive articulation due to readjustments at the end of the preceding period, arrive at a state of operations which are reversible and composable in well-defined groups. It is at this level that the notion of chance acquires a meaning as a noncomposable and irreversible reality antithetical to these operations. More precisely, it is at this level that the reality of chance is recognized as a fact and as not reducible to deductive operations.

The recognition of the existence of chance as well as its progressive interpretation as a nondeducible process are thus easily explained if we turn to the genesis of operations. In a general way, if we conceive of intuition as a representation of interiorized actions, operation derives from intuition as soon as this intuition becomes reversible, that is, as soon as the relationships at play are sufficiently determined so that it is possible to invert them. Reciprocally, the idea of chance is imposed when the conditions present remain indeterminate, which they can be from two distinct points of view: by comparison with spatial-temporal operations (physical chance) or by comparison with logical-arithmetical operations (drawing blindly from a group, or logical-arithmetical chance).

From the spatial-temporal point of view, the genesis of the notion of chance is particularly clear when we come to consider motion. Before seven to eight years, a displacement yields only a global intuition mixing dynamic factors with cinematic considerations. For instance, a child will deny that the sloping path AB moving uphill is equal to the path followed in the opposite direction BA. From the beginning of the second period (seven to eight years), the child is, on the other hand, capable of applying some operations to movement. For example, he succeeds in measuring the path taken as well as the duration and, without yet being able to generalize about the idea of speed, he is able to compare the speed of synchronous movements (in whole or in part) thanks to paths of differing lengths covered in equal times, or vice versa. Since he is in possession of these operative relations (not yet acquired during the preceding period, in spite of their apparent simplicity), the subject will understand from the beginning that the movement of a ball, which is determined when the trajectory is regular, escapes any possibility of deduction if collisions with other balls modify the trajectory, the speed, or the duration. At the intuitive level this indetermination posed no problem because the regular movements themselves remained indeterminate in the eyes of the children, and because the collisions and interactions which modified the multiple conditions remained unperceived. However, as soon as the spatial-temporal operations are developed, indetermination takes on a significance for the subject who has become capable of deduction as a function of the determining factors and leads him to recognize the existence of a domain apart: the domain of fortuitous events, unpredictable in that they are insufficiently determined or are indeterminable.

In short, to the extent that certain causal series give one a grasp of deductive considerations, physical chance comes up in the form, so well described by Cournot, as the interaction of independent causal series. The random mixture of marbles in a tipping box (Chapter I) is in this regard a typical example of overlapping (*imbrication*) between movements, each of which taken alone would allow a possible deduction, but which is rendered indeterminate because of the interaction among the parts. The rolling of metal balls down an inclined plane (Chapter II) is of the same type. In the case of the rotating needle (Chapter III), it is the friction at the pivot and the air currents which make the stopping point indeterminate, while the subject would like to consider the speed simply according to the initial push. As soon as several factors interact, the child who has become capable of deduction takes into account indetermination, and this discovery is then the source of the idea of chance. In fact, the indeterminate movements are then conceived of as noncomposable and irreversible, briefly as irreducible to operative grouping and as constituting, therefore, a domain apart, the uncertain.

From the point of view of logical or numerical ensembles, the situation is exactly the same. When the operations whose workings the child has just discovered are put into effect as a function of sufficient data and according to a certain logical order (that is, according to the laws of their groups), they lead to conclusions which the subject recognizes as necessary. But when the operations at play remain in part indeterminate, because sufficient data are lacking, or a new operation intervenes and interferes with the operations of a grouping with which it does not have a necessary link, there is a resulting indetermination. That is, there is the possibility of obtaining different results which lead the subject to consider that obtaining one of them will be a fortuitous event. The best example is the one using three counters, one of which is white (A) and two are red (A′) and which together form a set (B). While the child of the preceding level was not capable of including A and A′ in a permanent totality B, the subject of the present period knows quite well how to unite A and A′ in the form $A \cup A' = B$. But it follows that to think of A or A′, he finds that he is obliged (and it is here that the newness of the operative thought resides as distinct from intuitive thought), to proceed by inverse operations: $A = B - A'$ and $A' = B - A$. Now we ask him to choose an element belonging to B, with no further indication about this element. We have here either a case of an indeterminate operation or a case of a new operation, foreign to the grouping of the union of classes and meaning that he is to extract from the whole, B, one or the other of its parts. This operation will, in fact, no longer consist of drawing deductively an element from B according to one of the operations of the group (for example, $A = B - A'$), but to proceed by drawing from the midst of B, the result of the drawing remaining then indeterminate, that is, uncertain. While the determination of one of the terms of the group is possible only starting from at least two of the other terms (for example, $B = A \cup A'$; $A = B - A'$; $A' = B - A$; etc.), the drawing of any element from B does not fulfill this condition. This is why we must conceive of this operation neither as an insufficiently determined operation of the grouping of addition and of subtraction of classes nor as a new operation causing an interference with the operations proper to this system.

The result is that, from the construction of the concrete logical-arithmetical operations, a fundamental differentiation is introduced between the two modalities: the necessary (or the operatively deduced) and the possible (or indeterminate), without speaking of the actual data. This distinction of the necessary and the possible is explicitly achieved when concrete operations are translated into formal operations (or interpropositional ones) which the child of this level does not yet know how to do. But, as we have already seen, he has already succeeded in drawing from the union of sets, $A \cup A' = B$, which can be called a concrete disjunction: "If x is a member of B, it is then either in A or in A'." This nascent disjunction in its concrete form implies both the necessary ("if x is in B, it is necessarily in A or in A' ") and the possible ("if x is in B, it can be in A, but it is also possible that it is in A' "). Certainly this does not mean that this concrete disjunction is the same as the disjunction of the random drawings which we spoke of above, since disjunction unites in a single set two possibilities (as the union of sets unites into one set the whole of the parts), while a random drawing extracts a single element from a whole without supplementary determination. But disjunction, even when concrete, gives evidence of the duality of planes of the possible and the necessary, which remain unassociated in the union of sets.

In short, the discovery of indetermination (whether spatial-temporal or logical-arithmetical) which characterizes chance, by contrast with operative determination, entails the dissociation of two modalities, or planes of reality—the possible and the necessary—while on an intuitive level they remain undifferentiated in reality or in being. But then a new problem arises which is to dominate the entire spectrum of the second period and whose solution will be found only in the course of the third period: This is the question of the sample space, especially of the proportions of possibilities.

The sample space and proportions comprise two cases, the one additive and the other combinatoric. The first of the cases is presented, for example, when it is a question of extracting one or several elements from a set determined simply by a union of partial classes in a total class (a situation we have just seen in the drawing of one or several elements from a set $A \cup A' = B$). The sample space of possibilities is then directly furnished by the additive operation $A \cup A' = B$ determining the totality of possible cases. And the proportions of these possibilities (that is, the establishment of their probability) consist, consequently, of nothing more than establishing the relationship between the favorable cases and total number of possible cases: thus, for A, $A = B - A'$, and in the case of numerical data (for example, when A has one element and A′ has two elements), the probability of $A = 1/3$. On the other hand, it is possible that the sample space of possibilities is determined by the combinations, permutations, or arrangements between different elements, as, for example, when it is a question of drawing some pairs of elements from sets containing several given varieties with or without an order of drawing. The sample space as well as the proportions of possibilities arise from a second case with a multiplicative character which demands the use of combinatoric operations.

We note, therefore, that if the discovery of chance is due to the dissociation of the possible from the necessary, that is, the indeterminate from the operatively determined, the analysis of the possible supposes a reintroduction of the operation in the field of the uncertain. Thus, after a phase of clear opposition between the operative and the fortuitous, a new synthesis will sooner or later be effected between operations and chance, and this synthesis will characterize the notion of probability in a way distinct from the simple recognition of the fortuitous and the indeterminable. But this concept of probability supposes the analysis of all the possible cases as a total whole and not simply as isolated cases. Thus, it is only when the question of combinatoric analysis is resolved that this synthesis of operations and chance may be realized. Until then the antithesis remains irreducible because of the lack of operative composition of all the possibilities.

Given this situation, the second period of evolution which we now see is not only characterized by the discovery of chance due to the differentiation of the possible and the necessary, but also by the differentiation of the indeterminate and the operatively determined. It is characterized also by the absence of any probabilistic composition based on systematic and combinatoric analysis of the type that we will see in the third period. This is why the second period ordinarily begins with a skeptical phase in the course of which the child is limited to noting the unpredictability of isolated cases without thinking about or being able to consider all the possible cases together, the observed ones and especially the possible ones as the only source of probability. Next, he certainly succeeds in some cases in seeing the totality of the possible cases, but only those related to the first of the two varieties defined above, that is, the one concerning simply additive compositions as distinct from combinatoric compositions. The only judgments of probability—that is the only synthesis of chance with operations—which the child of the second period is capable of are, in fact, those which are based on simple static considerations of the union of parts of the whole as distinct from proportions and shifting combinations. It is for this reason that in the questions of Chapter VI the subject knows how to predict the probabilities (when there is a difference of the favorable cases with the same number of possible cases, or in the case of a difference in the possible cases). In the same way, in the case of unequal parts (in Chapter V, for example, one A, two B, four C, ten D), he knows how to judge them as a function of quantities. But already in this last example, when he has successful drawings, he no longer succeeds in taking into account the changing composition of the whole which is different after each drawing. Thus he considers only the initial composition and not the variable proportions, that is, the additive compositions and not yet the multiplicative ones (fractions and proportions).

In short, after distinguishing the possible from the necessary, the subjects of the second period fail to effect an exhaustive analysis of the possible. This again is easily explained as a function of operative development. Thought bearing on the possible is, in fact, formal thought which has precisely the characteristic of being hypothetical-deductive, that is, of operating on simple possibilities treated as hypotheses. It is thus natural that after discovering the possible, the child of the second period can at first think of it only intuitively, by analogy with the concrete operations which bear on the actual situation. At first he does not find a technique of reasoning adapted to this new modality any more than he at first found his operative technique in dealing with the actual situation. To get to this point he needed seven years of growth. In fact, formal thought is only developed between ten and twelve years. Therefore, it is only at this age that the child by using hypotheses as an intermediate step can apply deductive necessity to relationships intervening in the realm of the possible.

Now with chance (once the possible has been distinguished from the observed and from the necessary), it is precisely a question of describing the sample space, that is, all the possibilities, so that, on the one hand, the distribution of the total number of cases becomes predictable and, on the other hand, each isolated case acquires a probability expressed as a fraction of the whole. But to get to this stage, two conditions must first be fulfilled, both arising from what we have called earlier multiplicative compositions. In the first place, the subject must construct the combinatoric operations; in the second place, he must understand proportionalities, that is, he must succeed in seeing the relationships of two connected systems at once. Up to now we have defined formal thought by this capacity to reason about several systems at the same time—a capacity which is required also by combinatoric operations and proportions—while we are now characterizing formal thought as being the specific reasoning which bears on understanding probability. But we see immediately that these two characteristics of formal thought are united, because to reason on several systems simultaneously is the same as calling on the possible, and to reason on the possible means always to envisage several systems at the same time.

We understand the limits of the present period: In sum, if the subjects of this level already arrive at probabilistic judgments based solely on additive unions, they fall down when faced with any question needing proportions and combinatoric operations. In this regard, if we use again the nine points proposed for the first period, we then can conclude as follows:

In the first place, the idea of random mixture is globally acquired, but without an understanding of the multiple possibilities inherent in the trajectories according to the model of permutations and arrangements, and also without any understanding of the effects of large numbers (Chapter I).

In the second place, the irreversibility of uncertain mechanisms is understood, except again when it concerns large numbers.

In the third place, compensation between isolated cases ceases to be the undifferentiated type "each one gets a turn" and acquires a statistical meaning, but with the exception again of sets formed of large numbers (because of the failure to understand the proportionality of the differences).

In the fourth place, a beginning of symmetry is observed in the centered distributions (bell curve type), but still only for small numbers.

In the fifth and sixth places, predictions of a purely empirical type diminish, while a beginning of inductive reasoning is shown whose function consists of distinguishing between regularities and fortuitous distributions.

In the seventh place, there is the consciousness of the miracle which an extremely unlikely distribution would mean, and an inductive search for a cause to explain such distributions other than by chance (for example, the fixed counters of Chapter IV).

In the eighth place, the simple operative unions make possible all judgments of probability based on the relationship of the whole with the part when one of these two terms remains invariable. On the other hand, the comparison of two proportional groups, or groups where the whole and the part vary simultaneously, still remains impossible.

Finally, in the ninth place, there is not yet any ability to interrelate the whole of a field of distributions, notably in the case of large numbers.

These traits of the second period, therefore, form again a homogeneous whole entirely explainable by operative development: The appearance of concrete operations explains the discovery of chance by antithesis with deductive necessity, while the absence of formal operations explains the lack of a synthesis between chance and the operative mechanisms in the form of a system of probabilistic composition.

4. *The third period: Probabilistic composition, a synthesis of chance and the deductive operations.*

As soon as chance is discovered as indeterminate relationships not composable by operative methods and, contrary to all the operations, irreversible, the mind seeks to assimilate this unexpected obstacle encountered on the road to developing deduction. There is simply one manner of understanding and explaining this: Group the relations at play according to the model of the operative systems. If chance for a moment makes reason useless, sooner or later reason reacts by interpreting chance, and the only way of doing this is to treat it as if it were, in part at least, composable and reversible, that is, as if one could try to determine it in spite of everything.

It is from this need that probabilistic composition is born, already sketched out since the beginnings of the second period, but not to be achieved until the third period. The principle is simple: If isolated or individual cases are indeterminate or unpredictable, the whole as such is, on the contrary, determinable by a system of multiple possibilities, and composition then will consist of tying together the parts again (individual observed cases) with the whole thus defined (all the possible cases).

This is what the child understands from the beginning of the second period in the cases where the whole is formed by the simple addition of a few elements. If $B = A \cup A'$, for example, we do not know if we will draw A or A′, but we are certain to draw one or the other, since we are drawing from group B. But the problem is much more general as soon as we propose to make predictions not simply on whether we will draw a single element of a set of mixed elements, but are drawing two elements or three, or even predicting the order of drawing, or the order and the number predicted together, and so on. It is here that combinatoric analysis and a consideration of proportions necessarily intervene.

There is no need for us to retrace here the genesis of combinatoric operations studied in Chapters VII to IX, nor the genesis of proportions studied elsewhere,[1] for neither of them is developed necessarily as a function of chance. Our present problem is, on the other hand, to understand how these operations, once they have been acquired, react on chance, that is, how the subject, in order to assimilate chance with a play of operations, will use the operations so as to understand relative determination which constitutes probability.

These things are easily understood. Their principle consists simply (as we have just hinted) in treating the uncertain mechanisms as if they were not uncertain, that is, as if they were no more

[1] Jean Piaget and Bärbel Inhelder, *The Child's Conception of Space* (New York: Norton, 1967), chap. XII.

than the result of transformations, each of which taken alone, could be operative. The most striking case is the one of random mixture. Nothing is more unpredictable and indeterminate than the innumerable collisions possible when twelve marbles hit each other in the course of the movement started by the tipping of the box, and with trajectories determined by the collisions. Nevertheless, once the child has learned the operations of permutations, he does not hesitate to consider the mixture as if it were the result of an action consisting simply of permutating the dozen elements in all the possible ways, just as though he could do it himself and just as he does, in fact, achieve it in his drawings of the trajectories. But since chance is not an operative system, we understand that the invisible permutations remain fortuitous, that is, (a) instead of being effected according to a systematic order, they move pell-mell in all directions, and especially (b) instead of being complete, they are able to achieve only certain of the possibilities. However, the results can be compared: While systematic calculation furnishes all the possibilities simultaneously, chance realizes only certain possibilities, but creates nothing new and remains necessarily within the framework of deduced possibilities, its only originality is being disordered and incomplete. In the same way, the drawing of pairs of marbles from a sack containing several colors (Chapter V) can be done according to all the combinations of pairs: Operations predict all the possibilities, some of them coming about by chance, which are thus observed ones, but they also must be among the original possible cases. In short, instead of letting chance keep its unpredictable character, incomprehensible as such, the mind translates it into the form of a system of operations, which are incomplete and effected with no order. Chance subsists, therefore, but has been moved to the plane of operations where it gains intelligibility. Chance ceases to be uncomposable, to speak in an absolute way, it is assimilated with a group of operations, themselves effected according to chance and in restricted number. Thus chance becomes comparable to those very operations conducted systematically and in a complete manner.

It is starting from this point that specifically probabilistic composition is developed. The crucial example is again the one of random mixture, since a mixture is an irreversible process, while the system of operations of permutations to which it is related by analogy is, on the contrary, entirely reversible to the extent that the permutations form a typical mathematical group. Children of the third level furnish, on this point, perfectly relevant observations: The actual random mixture can come back to its point of departure (what the little children call an unmixing), but to achieve this, we need a very large number of trials because *"that's very rare," "there is almost no chance of that,"* and so forth. In other words, the operative construction which permits describing the sample space of all the possibilities leads by the same process to a conception of the relationship between the favorable cases and the sum of all the possible cases.

This judgment of probability is thus only the result of a comparison between chance, irreversible and uncomposable, because it incompletely realizes the possibilities and the composable and reversible operations with which it can be partially assimilated. Judgments of probability are then certainly a synthesis—partial but pushed as far as possible—between chance and operations. The operations lead to the determination of all the possible cases, even though each of them remains indeterminate for its particular realization. Probability then consists in judging isolated cases by comparison with the whole, which is, therefore, in part determined and confers on the isolated cases a fractional coefficient of realization. Probability, therefore, conforms with its actual definition, a fraction of determination.

These mechanisms explain the double progress which characterizes the third period: structuring of the totalities leading to the law of large numbers, and the probability of isolated cases as a function of the whole. These are the two linked traits which constitute probabilistic composition.

Concerning the first point, there is a very remarkable reaction which at first contrasts the subjects of the third level to those of the second: Faced with a problem either of physical chance (random mixture, the rolling of seeds from a funnel down an inclined plane, rotation of a needle, etc.) or of drawing random lots, the child of the second period begins with an attitude of general doubt about predictions (*"it's chance,"* thus we cannot predict). This is due to the antithesis just discovered between chance and operations.

At the third level the child seeks at first to see and structure the whole distribution (the whole of the possible permutations in the case of random mixture; bell-shaped distribution in the shape of Gauss's curve in the case of the seeds; uniform distribution of the possible stopping points for the rotating needle; quantitative distribution of the values in play with the random drawing, etc.).

This structuring of the whole, or the discovery of the collective distribution, is psychologically the true and single basis for probability. When the subjects of the second level make judgments of probability, it is already by means of an intuitive anticipation of the whole distribution (anticipation based at that time on past experiences and not yet on a combinatoric scheme) that they are making such decisions. It is these intuitions that find, on the third level, their equilibrium in the operative construction of the whole, and it is only at that time that the mechanism of reasoning becomes visible. This structure of the whole is extended as a kind of judgment of probability, the touchstone we could say of the existence of a probabilistic composition, which is the understanding of the law of large numbers. The subjects of the second level already discover as soon as they anticipate a probability intuitively what we have called the law of small large numbers, but they are not able to generalize beyond twenty or thirty elements. The true law of large numbers marks, on the contrary, the point of equilibrium of probabilistic composition. This law means simply, as we know, that the limit of relative frequency equals the probability. When the subject understands the law of large numbers it demonstrates clearly that he has a good grasp on both the probabilistic nature and the combinatoric composition possible in the distribution of the whole. When Lau (13; 0), for example, in working with the rotating needle (Chapter III), says that by repeating the experiment *"it will always become more equal,"* this is clearly a sign that he is judging that there will be an equal distribution on each of the colors on which the needle stops and also that if each stopping taken alone is fortuitous, the distribution of the whole, on the contrary, is itself deducible.

By strict correlation with this analysis, the probability of isolated cases does then take on a general meaning which it was still lacking in the course of the second period. Up until now, in fact, the prediction of individual cases was based simply on an examination of previous drawings with the intervention of a compensatory element destined to anticipate a more or less regular distribution. But the problem was then to reconcile these successive intuitions with the famous principle of Joseph Bertrand, "chance has neither consciousness nor memory," which was suspected as soon as the idea of chance was imposed on the mind of the child. However, with the structuring of the distribution of the whole, an isolated case acquires the precise meaning of a fraction of the whole, and is no longer tied to what he observes, but is related to all the possible cases which are realized little by little as a function of the law of large numbers.

In addition and going right along with this, at the level when probabilities are susceptible to such propositions, inductive reasoning becomes, for that very reason, systematic. Since a real fact is always a composite of chance and of transformations expressable in terms of determined operations, inductive reasoning is then what delimits the two domains and combines them according to their respective parts. In the bell-shaped distribution (Chapter II), for example, induction takes into account the rollings of balls from the funnel mouth while setting aside the statistical symmetrical distribution from the point of maximum frequency. However, in the case of the magnets (Chapter III), induction focuses on the role of this symmetrical factor in contrast with the distribution of the fortuitous stopping points.

Thus we see what probabilistic composition consists of in the third period. It is a synthesis of formal operations bearing on the possible, and of chance itself, and recognizes the uncomposable and irreversible character of chance while reserving for combinatoric operations, which are composable and reversible, the role of an entirely deductive determination of the possibilities. These possibilities are realized only at the limit which is marked by large numbers, while effective chance remains indeterminate in different degrees, each individual case being only a fraction of determination, whether it remains entirely fortuitous (the determination then being reduced to a comparison of the case envisaged with all the possible cases) or it is partially fortuitous (i.e., there is then inductive determination by the interaction of chance and operative transformations).

5. *Chance and probabilities.*

The psychological study of the genesis of the notions of chance and probability carries a certain number of lessons which are worth examining from the point of view of the psychology of operative thought in general.

Once reasoning has been acquired, that is, reasoning as it is presented in balanced states of thought, it is always deductive, and deduction rests on systems of logical-arithmetical or spatial-temporal operations which consist partially of groups and groupings. It is with reference, and more precisely, by antithesis with these group operations that the idea of chance is imposed on the intellect. By the word *chance* we mean relatively uncomposable and irreversible processes. In contrast with deductive determinations, chance is, therefore, indeterminate, and this relative indetermination can be found in three forms.

1. There can be insufficient determination, from the point of view of logical-arithmetical operations, when certain new operations interfere with the order of construction according to which the basic operations are generated. We have seen this in an example with a logical characteristic when the new operation consisted in choosing from B any element when the basic operations which form B are an addition of the classes $A \cup A' = B$. A and A′ are found by means of subtractions, $B - A' = A$, and $B - A = A'$ (operations with at least three terms). In this case the element x to be chosen in some way from B remains indeterminate since it can be in A or in A′. A well-known example of interference of the same type is found in the seventh decimal place of logarithms: If we draw at random the numerals given, for example, in the seventh decimal place of natural logarithms, we can find for the ten possibiilties 0, 1, 2, . . . 9, a frequency approximately equal to $\frac{1}{10}$ for each digit. However, nothing is better determined than

the seventh decimal place of a particular logarithm. However, it is clear that this seventh decimal is determined only from the point of view of the operation which generates logarithms (a correspondence between arithmetic and geometric progressions). If, on the other hand, a new operation interacts, one which consists of choosing a logarithm at random, each time extracting the numeral in the seventh decimal place, there is then an interference between two sorts of independent operations (as, a moment ago, by choosing any element of set B without going through the unions or complements of sets). From the point of view of this new operation, the numerals in the seventh decimal place are indeterminate. That is, we get the uncertain distribution of numerals which compose them, which certainly constitutes, therefore, an effect of chance.

In a general manner, chance due to the interference of independent logical-arithmetical operations is reduced, therefore, to a logical indetermination which can be expressed by the notion of nonimplication or, more precisely, of deductive noncomposability. From this point of view, every quality attributed to a logical entity can be considered as fortuitous without its being implied by it. For example, the concept of swan implies the qualities of a vertebrate, of a bird, and so on, but whiteness is for the swan a fortuitous quality since there are black swans. From the operative point of view we can even say, no doubt, that the intransitive character of certain relations (as different from the transitivity and the alio-transitivity [1] permitting the grouping) expresses one part of chance: The fact that "A is a friend of B and B is a friend of C, but A is not a friend of C" expresses a fortuitous factor in the formation of friendships, while "A equals B, B equals C, therefore, A equals C" expresses a deductive order foreign to chance.

2. On the other hand, there is chance in cases of insufficient determination from the point of view of spatial-temporal or physical operations. At this point, it is no longer a question of interferences between two series of related operations, but between two independent causal series, and we find then the definition of chance given by Cournot. We must also remark that this new case is exactly parallel to the preceding one with the single difference that

[1] We call alio-transitivity the multivocal transitivity of certain relations, for example, the brother of my brother may be my brother or myself.

chance here is no longer reduced to logical nonimplication or noncomposability but to noncausality which is true in the following sense: Causality proper to the series A (for example, a man walking down the street) does not necessarily include the interference with the series of causes and effects B (for example, the falling of a tile due to wind), and reciprocally.

3. Finally, there can be chance by double indetermination which is at once logical-arithmetical and spatial-temporal. The threshold of indetermination which, according to Planck, marks the limit of the microphysical compared to the macrophysical is, no doubt, characteristic of this third type. Below this threshold the notion of object loses its meaning, while in the domain of phenomena in our scale, the object is at the same time the point of departure of logical-arithmetical operations (a set is a collection of objects) and the point of arrival of spatial-temporal operations (which bear on the composition of the object itself). We must simply note that in the case of double logical-arithmetical and spatial-temporal indetermination, chance can no longer be conceived of as due to the interference of independent logical or causal series, each of which would be quite well determined in its own right. Instead, at this point, it derives from a more basic indetermination which is due to the limits of physical action and to the logical-mathematical operations themselves which converge with the limits of the object.

But, if in these three sorts of cases, chance is reduced to indetermination, indetermination is doubly relative. First of all, it is relative in the sense that indetermination is never complete since the consideration of distributions permits one to attribute to each event a certain coefficient of probability, thus a fraction of determination. But it is also relative to a second point of view in the sense that thought meets the indeterminate only in relation to a considered system of operations. We, therefore, never know if a fortuitous event is in itself indeterminate or only by comparison with the operations at play, and psychologically only this relationship counts. It is in this way that the example of the seventh decimal of the logarithms shows rather well the relative character of chance which intervenes in a particular case, especially since this decimal place is well determined. However, philosophers have at times stretched the notion of indetermination to the absolute. Mr. A. Reymond held, for example, that all judgments of probability rest on the equal likelihood of all possible cases. This equality of possibilities is a real contingency as opposed to determinism[1] (even macroscopic). However, we have some difficulty in following this argument because equivalence of possibility is itself always relative to the proposed operation. On the other hand, the mathematicians no longer base the calculation of probabilities on the equal likelihood notion and start directly from the observation (or hypothesis) of a distribution. As for the threshold of indetermination of the microscopic, we have naturally sought to interpret it also in the sense of a realistic antideterminism (which would tend to justify the demonstration due to physicists themselves according to which microscopic indetermination could not include any underlying determination). And from the point of view of an operative psychology of thought, it remains no less true that the threshold of indetermination proves uniquely the insufficiency of our usual operations and of our macroscopic representative schemes to insure the intelligibility of the phenomena when the limits of a certain scale of action are exceeded. It is true, therefore, that chance and microphysical indetermination teach us the limits of our effective activity and the intellectual effects of this limitation of action. Therefore, it is proper to await the construction of new processes of thought adapted to the new scale rather than to anticipate the perpetual inadequacy of our instruments of intelligence.

From the psychological point of view, indetermination which is proper to chance is reduced to a relative independence of the possible operations on the specific objects and to the limits of our usual operations on objects in general. This relative independence, as well as these shifting limits, demonstrate more certainly the character of progressive adaptation of thought (obliged to divide domains so as to conquer them separately), than they do the contingency or noncontingency of a reality envisaged in itself. From the experimental reality of physical chance, psychology, therefore, invites us to derive an epistemology rather than a metaphysics.[2]

[1] A. Reymond, *Philosophie spiritualiste* I: 364–66.

[2] See our *Introduction à l'épistémologie génétique,* chap. VI. (Paris: Presses Universitaires de France, 1950).

Under this same psychological perspective, the intervention of chance in the cognitive processes presents, on the other hand, a decisive importance concerning the limits of operative activity since chance is always, in part, related to this activity. Uncomposable and irreversible, chance constitutes the pole opposed to the pole of grouped connections. However, by this very fact, it defines a kind of reality in which the whole is no longer equal to the sum of the parts, because of the lack of additive composition, but obeys its own laws which are laws of equilibrium *sui generis*. From the point of view of the object it is in this way that a mixture is uncomposable in terms of additive composition from the very fact of the multiple interferences which characterize it. It does, however, obey the laws of the structure of the whole based on considerations of equilibrium, equilibrium being oriented in this case in the direction of the most probable combinations. Contrary then to the rules of mechanical composition in which a system is equal to the sum of its parts, statistical equilibrium supposes a continual modification of the parts as function of the whole, precisely because chance is present. For example, the shifting about of heat in a closed system depends on the total thermic equilibrium, the intervention of a new body being able to modify the whole of the other parts without the heat of any one of them remaining identical to what it was before. From the point of view of the subject, the same is equally true since chance intervenes in the mental processes in the same way as in the material world. In the same way, in fact, as the indetermination of operations applied to the object expresses a physical chance under the species in which we conceive certain material realities, so also the indetermination of operations envisaged from the point of view of the subject (in so far as the subject does not succeed in coordinating them because he lacks a sufficient operative mechanism) indicates interior or mental chance. It is all the more necessary to consider this latter fact because the functions in play are more elementary and further removed from rational intelligence. Chance which is proper to psychic phenomena is due to the fact that consciousness can never bring together all the data into one united field except precisely in the logical or deductive domain. Consciousness proceeds then by successive centrations around a center motivated by the interest of the moment which stimulates them, but more or less fortuitous in the points of their application and, therefore, intersecting the data at more or less uncertain points. Alongside the reversible mechanisms structured according to the laws of groups, such as we find in the mechanisms of intelligence once it has arrived at its state of mobile equilibrium, there is the intervention in mental activities of the irreversible and operatively noncomposable mechanisms such as associations, motor habits, and so on. In these mechanisms also, the whole is something else than the sum of the parts. Such, in particular, are the perceptive mechanisms which obey the laws of *Gestalt* and not those of grouping. As we have tried to show by analyzing a certain number of perceptive structures and especially the illusions deriving from Weber's law, it is precisely because these perceptive structures carry a part of chance and obey essentially statistical laws that the perceptive mechanisms present forms which are irreducible to additive composition and, consequently, to operative intelligence as well.

Remaining within the confines of the intelligence, the conflict between irreversible chance and the operative reversible mechanisms ends up finally in a translation of chance itself into the language of operations. This is what we have just noted by studying the genesis of probabilistic composition, a synthesis of chance and the combinatoric operations. This process of thought is even more general and has a bearing on all the irreversible systems. It is thus that in physics the thermodynamic phenomena which are essentially irreversible can be interpreted only by reference to reversible mechanisms and they are themselves translated in terms of reversibility. We can imagine, for example, processes happening at an infinite speed and constituting a continuous series of states of equilibrium which, in fact, is impossible, but which allows us to describe an irreversible transformation (with, naturally, a finite speed) in the language of reversibility. Similarly, in microphysics, however primitive and essential uncertainty may be considered in opposition to determination, the calculation of the states and the transformations is at first tied to systems of operators. These systems introduce the reversibility of the actions of the observer into the midst of the most irreversible objects observed, even to the point of creating the same symbiosis between the operative composition and the actual as in thermodynamics, but built according to this new scale on a fundamental principle of physical knowledge.

It must, however, be well understood that what we call in this way probabilistic composition is in no way to be confused with the calculation of probabilities in general. We must, in fact, carefully distinguish the mathematical or purely operative theory of probability and its applications which alone constitutes what we designate here under the name of probabilistic composition.

The mathematical theory of probability is a pure deduction resting on an axiomatic structure, as does every pure theory, and thus constitutes a system composed exclusively of formal or hypothetical-deductive operations. We will define, for example, the function of probability of uncertain variables between minus x and plus x, and we will study the different combinations of such functions. To the extent that probability is a purely operative system, it will derive from additive composition (cf. the theory of total probabilities where the notion of the addition of two probabilities is introduced, etc.), and the remarks made above concerning the relationships of the whole to the part in physical systems implying chance are in no way, therefore, applicable to the mathematical theory of probability. But the counterpart of this deductive purity is that the theory thus constructed deals only with the possible, what mathematicians express as abstract.

The theory of probability applied to the real world presents, on the other hand, the essential characteristic of always containing a scale of approximation, that is, of neglecting probabilities which are too small. This is equivalent to saying that in passing from the total system of cases theoretically possible—which comes from a completely additive composition—to the system of cases actually able to be realized, the field of possibility itself is restricted, and hence the abandonment of the notion of additive totality. M. Emile Borel teaches, for example, that on the human scale negligible probability can be considered as being at 10^{-6}, on the terrestrial level, 10^{-15}, and on the cosmic level, 10^{-50}. This means that if the probability that an event not happen is less than 10^{-50}, we can exclude it from the field of reality.[1] It is precisely by neglecting

[1] M. Emile Borel, *Valeur pratique et philosophie des probabilités* (Paris: Gauthier Villars, 1939), pp. 6–7.

these extremely improbable events that one is assured physically of the irreversibility of a random mixture, while in a mathematical sense the return to the starting point is always possible. It is thus that from the physical point of view, the growth of entropy (the principle of Carnot-Clausius) is the model of necessary irreversible evolution, while mathematically the movement of heat from a colder body to a warmer one does have an assignable probability. It is thus exclusively from the point of view of probabilities applied to the real world that we can set up the opposition between probabilistic composition and operative composition, and seek the criterion of the former in the fact that the whole presents more traits than those of the added parts. Since only the most probable combinations are realized (with exclusion of the combinations lying below the scale of approximation), the system of real interactions is irreversible even though the combinatoric operations permit an analysis of it and allow a determination of the probabilities at play and are themselves reversible.

This brings us to the question of the value of the probability of an isolated event, a problem which is no doubt at the center of the theory of applied probability. It seems that psychological analyses can help us solve this question by identifying what is implicitly contained in the most elementary probabilistic intuitions.

Certain axiomatic or logical works on the calculation of probabilities such as those by von Mises and Reichenbach have tended to show that all probability is necessarily relative to a collection of events and, therefore, to a certain distribution or to a system of frequencies posed at the start. Now not only do such conceptions allow us to avoid the pseudo problem of equivalence of possibilities, but also they tend to put into doubt the value of the notion of the probability of an isolated event. On the other hand, M. E. Borel seeks to preserve this notion by invoking a certain number of cases in which judgment of probability seems to him independent of preceding statistics. Examples of this are the judgment of a doctor on the probable death of a patient with tuberculosis who presents a rare form of the disease such that the class "would be reduced to the limit of a unique case," [1] or the judgment expressing the

[1] *Ibid.*, p. 88.

probability of the weight of an ordinary suitcase, that is, more than one kilo and less than fifty kilos.[1]

From the psychological point of view it seems quite difficult not to see in every judgment of probability a reference, whether implicit or explicit, to a system of distributions or frequencies. In fact, we have noted that the intuition of probability was in no way primitive for the child and that it began only from the moment when this consideration of the entire distribution was developed. During the first period of this development, that is before seven or eight years, it is true that there were some judgments bearing on isolated cases independently of the consideration of frequencies.[2] But we could not, in this regard, speak of their judgments of probability, since they come before any notion of chance and since they make decisions about the occurrence of a future event in an arbitrary or subjective manner, not based on probabilistic considerations. As soon as the notion of chance appears (at the beginning of the second period), the child begins by denying the possibility of these isolated predictions to which he had earlier accorded an illusory value. Then he regains his confidence, but this time with motivation: The true intuitions of probability begin only then, but precisely as a functon of an operative synthesis which subordinates the isolated prediction to the consideration of a whole system of frequencies. The development of a system of distributions is, therefore, certainly the pre-existing psychological condition of probabilistic intuitions.

If, from this point of view, we take another look at the examples advanced by M. Emile Borel, it seems clear that they make the same appeal to a totality of unconscious comparisons with previous knowledge, thus to an implicit table of experimental frequencies, however rudimentary they may be in their mathematical development. In a case of tuberculosis truly unique in its class, the doctor can predict nothing. If he profers a judgment of probability, risking a diagnosis, it is surely by analogy with certain traits of other cases, even if partly different, that is, by the union of the singular

[1] *Ibid.*, p. 93.

[2] Except when concerned with the precocious idea of compensation which, in fact, implies that the subject does possess a certain idea of frequency even though his interpretation is subjective.

case under consideration with more general ones. As for the weight of a suitcase, it is difficult to believe that an individual certain of the limits more than one kilo and less than fifty kilos has not had the occasion of making previous experiments with such a situation. A table of distributions is, therefore, almost certainly inscribed in his practical habits, even if it includes only quantifications such as, often, ordinarily, rarely, almost always, or almost never, and so on.

The question of isolated probability is, however, more complex than these banal reflections would allow us to suppose. There are, in fact, two methods of analyzing a new and singular case. One is by referring it to previous experiences, and thus to explicit or implicit statistics of analogous cases. But the second method consists of a logical or mathematical construction of the relationships implied in the case to be evaluated. In this way M. Darmois [1] describes this second analytical procedure by distinguishing the following three steps: (a) a schematization appropriate for the phenomenon to be studied; (b) an anaylsis of the possibilities leading to a construction of all the possibilities in play; and (c) a consideration of this set of all the possibilities. The true difference between these two methods consists in seeing that, in the first one, the entire set of the frequencies is given ahead of time and in extension, while in the second one, the set of frequencies is constructed in the light of the new case as it is analyzed in intension.[2] But in both situations, the reference to a display of all the possibilities is necessary because we could neither schematize a new phenomenon nor establish a qualitative or quantitative table of the possibilities from which it derives without basing ourselves on a group of previously given schemes, more or less general, and themselves supposing a system of frequencies.

In short, the conflict of the isolated probability and the primacy of the notion of distribution recalls the logical conflict of concepts in intension and classes in extension. Now, in the same way that the relationships in intension given between the defining traits of an individual are always correlated to the unions in extension

[1] In a note from the book by M. E. Borel already cited, pp. 162–66.

[2] *Extension* and *intension* are terms of formal logic. Extension refers to the classes of items to which a term applies; intension refers to the meaning of a term (translators' note).

which permit us to situate this individual in a total system of classes; so also judgments of probability which we bring to bear on what are in appearance the most isolated or singular events are always related to the classes in extension by means of which we succeed in comparing these events with others (analogous or different), and thus are always dependent on a table of distribution. In a word, in the same way that the necessary in intension corresponds to always or to all in extension, so also the different degrees of the probable in intension correspond bi-univocally to the different values of the frequency in extension.

Bibliography

Major related works by Piaget and his collaborators

Inhelder, Bärbel, and Jean Piaget. *The Early Growth of Logic in the Child.* New York: Norton, 1969.

———. *The Growth of Logical Thinking from Childhood to Adolescence.* New York: Basic Books, 1958.

Piaget, Jean. *The Child's Conception of Movement and Speed.* New York: Basic Books, 1969.

———. *The Child's Conception of Number.* New York: Norton, 1965.

———. *The Child's Conception of Time.* New York: Basic Books, 1970.

———. *Genetic Epistomology.* New York: Norton, 1972.

———. *The Language and Thought of the Child.* New York: Harcourt, Brace, 1926.

———. *Judgment and Reasoning in the Child.* New York: Harcourt, Brace, 1928.

———. *The Moral Judgment of the Child.* New York: Harcourt, Brace, 1932.

———. *The Origins of Intelligence in Children.* New York: Norton, 1963.

———. *Play, Dreams, and Imitation in Childhood.* New York: Norton, 1962.

———. *Understanding Causality.* New York: Norton, 1974.

———, and Bärbel Inhelder. *The Child's Conception of Space*. New York: Norton, 1967.

———, Bärbel Inhelder, and Alina Szeminska. *The Child's Conception of Geometry*. New York: Basic Books, 1960.

Glossary

The following terms represent key concepts or technical aspects in the works by Piaget; the brief definitions and examples given here may prove useful to the reader facing Piaget for the first time.

CENTRATION This is a characteristic of preoperational thought in the child that causes the child to center on one factor of what he or she observes to the exclusion of other, equally important factors. For example, when a child sees a fixed volume of liquid poured from a short, wide container into a tall, thin container, the child might think the quantity has increased because he or she centers on the height of the new container and ignores the width.

DECENTRATION This represents the ability to consider several aspects of a situation simultaneously, that is, to avoid centration.

GROUP This is a mathematical term. A group is an algebraic structure consisting of elements and an operation. With respect to the elements, the operation is closed and associative; there is an identity element and each element has an inverse element. A simple example is the set of integers and addition. Addition is closed on the set of integers since the sum of any two integers is another integer. Addition is associative since for all integers $(a + b) + c = a + (b + c)$. Zero is the identity element since for all integers $a + 0 = a$. Every integer has an inverse (called its negative) so that for all integers $a + (-a) = 0$. The importance of this term for Piaget is that he uses mathematical models to describe cognitive structures, and a group is one of the most important mathematical structures.

GROUPING This is a hybrid mathematical model or structure originated by Piaget evolving from the two mathematical structures known as a group and a lattice. In Piaget's theories he uses groupings as basic models for cognitive structures. A grouping is, therefore, more complicated than a group since it incorporates the lattice concept as well as the group concept. (A lattice [1] is formally defined as a partially ordered set L in which each pair of elements has a greatest lower bound and a least upper bound. Unfortunately, this definition is of very limited value, except to someone trained in higher mathematics, since a lattice is an abstract mathematical structure and the definition itself contains several technical terms. Perhaps it will help to say that an example of a lattice is the set of all subsets of some universal set, where the partial ordering is the subset idea; the greatest lower bound of any two subsets, A and B, is $A \cap B$, and the least upper bound is $A \cup B$.)

OPERATION This is perhaps one of the most basic concepts in the works of Piaget, a concept which he further distinguishes into formal operations and concrete operations. First, by operation in general Piaget means cognitive actions which are organized closely together into a strong structure. For example, the sorting of a set of objects into subsets with common characteristics (color, size, shape, etc.) is one kind of operation. The union or intersection of sets are other operations, and so are the usual operations of ordinary arithmetic. Flavell says "any representational act which is an integral part of an organized network of related acts is an operation." [2]

Concrete operations are those performed by a child on what is immediately present or observable. They refer, indeed, to the concrete or the actual.

Formal operations, on the other hand, deal with the possible or potential, not just the real or actual. They are concerned with hy-

[1] See Simmons, George F., *Introduction to Topology and Modern Analysis* (New York: McGraw-Hill, 1963).

[2] John H. Flavell, *The Developmental Psychology of Jean Piaget* (Princeton, N.J.: Van Nostrand, 1963), p. 166.

pothetical-deductive thought which is propositional in nature. They are operations on operations or, as Piaget sometimes says, "operations to the second power."

REVERSIBLE This term refers to a property of operations. An operation which is reversible has an inverse (in the mathematical sense of a group). For example, multiplying a number by 5 is reversible since the result can be undone by multiplying by $\frac{1}{5}$, the multiplicative inverse of 5. Another example of reversible operation is the cutting of a ball of clay into several parts. This is reversible since the parts can be stuck back together to restore the original ball. On the other hand, some actions are not reversible, such as the burning of a piece of paper or the scattering by chance of a handful of grass seed.